Following Nellie Bly

Nellie Bly, circa 1890. (*Courtesy of the Library of Congress, LC-USZ62-75620*)

Following Nellie Bly

Her Record–Breaking Race Around the World

Rosemary J. Brown

SS *Augusta Victoria*

PEN & SWORD HISTORY

First published in Great Britain in 2021 by
Pen & Sword History
An imprint of
Pen & Sword Books Ltd
Yorkshire – Philadelphia

ISBN 978 1 52676 140 8

Typeset by Mac Style

Printed and bound by CPI Group (UK) Ltd, Croydon, CR0 4YY

Pen & Sword Books Limited incorporates the imprints of Atlas,
Archaeology, Aviation, Discovery, Family History, Fiction, History,
Maritime, Military, Military Classics, Politics, Select, Transport,
True Crime, Air World, Frontline Publishing, Leo Cooper, Remember
When, Seaforth Publishing, The Praetorian Press, Wharncliffe
Local History, Wharncliffe Transport, Wharncliffe True Crime
and White Owl.

For a complete list of Pen & Sword titles please contact

PEN & SWORD BOOKS LIMITED
47 Church Street, Barnsley, South Yorkshire, S70 2AS, England
E-mail: enquiries@pen-and-sword.co.uk
Website: www.pen-and-sword.co.uk

Or

PEN AND SWORD BOOKS
1950 Lawrence Rd, Havertown, PA 19083, USA
E-mail: Uspen-and-sword@casematepublishers.com
Website: www.penandswordbooks.com

For Pauline

Contents

List of Illustrations

Drawings by David Stanton

List of Plates

Front cover: Nellie Bly in travelling ensemble, Library of Congress LC–USZ62–59923

Back cover: Nellie Bly publicity photo for *The New York World*

1. Nellie Bly in her legendary travel attire.
2. 'Nellie Bly Bids Fogg Good Bye' trading card.
3. Author's daughter Acadia on the steps of the Maison Jules Verne in Amiens, France.
4. Grand Oriental Hotel, Colombo, Ceylon, circa 1890s.
5. Author at Full Moon Poya celebrations in Colombo, Ceylon.
6. The replica statue of Sir Stamford Raffles, Singapore.
7. Author touring temple in Singapore's Little India.
8. Hong Kong Harbour, 1889.
9. Happy Valley Cemetery swamped by skyscrapers, Hong Kong.
10. Firing of the Noonday Gun, Hong Kong.
11. 14th century Zhenai Tower, now Guangzhou Museum, Canton, China.
12. Temple of the 500 Gods, Canton, China, in Nellie's time.
13. Today's Temple of the 500 Gods, Canton, China, now called Hualin Temple.
14. Tea at the Grand Hotel, Yokohama, Japan.
15. Sangedatsumon Gate, Zojoji Temple, Tokyo, Japan.
16. Lap of the Great Buddha, Kamakura, Japan.
17. Nellie Bly's triumphant arrival and reception in Jersey City after circling the world in seventy-two days.
18. 'She's Broken Every Record!', front page of *The New York World*, 26 January 1890.
19. Nellie Bly's Round the World Game.
20. Staircase at the New York City insane asylum where Nellie Bly went undercover.

Chapter 1

In Which Nellie Proposes to Girdle the Earth

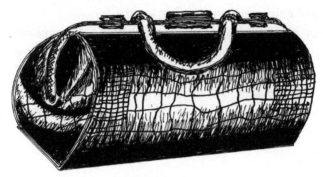

Nellie Bly's gripsack

New York City
November 1888

'If I could do it as quickly as Phileas Fogg did, I should go.'
Nellie Bly

A clock taunts the hours, shattering the early morning silence in the apartment that journalist Nellie Bly shares with her mother in Manhattan. All day long and into the night she has scoured her brain. Sunday is her customary day for preparing story proposals for her editor at *The New York World*. When she finally stumbles into bed, the prospects of finding an idea are as bleak as the bitter November darkness outside. Her quilt lies on the floor, sheets escape from her mattress, her nightclothes seem to strangle her. It is 2.45 am; Monday is already here and she has nothing, not a single proposal for her meeting later that day. Frazzled by fatigue and frustration, all she wants to do is escape. 'I wish I was at the other end of the earth!' she exclaims in the pre-dawn hours. Nellie stops to ponder her words, raising an eyebrow. 'And why not?' she asks herself. For two years she has worked non-stop at *The New York World*, the leading paper of its time. Migraine headaches are setting in. 'I need a vacation; why not take a trip around the world?' she says. One thought leads to another and by 3.00 am Nellie has a plan. She will challenge the record of Jules Verne's hero in *Around the World in 80 Days*. 'If I could do it as quickly as Phileas Fogg did, I should go.' With that Nellie drifts off to sleep, determined to discover that very day if she can circle the world in less than eighty days.

That was it! The idea, born of exhaustion, that ignited a record-breaking journey that would define Nellie's life, make her the most famous woman of her time, launch a legacy that lives on today, and make the world a little smaller. The idea that, 125 years later, would inspire me to pay tribute to Nellie by following her footsteps around the world.

*   *   *

Like Nellie, I like to set aside Sundays, usually after lunch, to pursue ideas, not so much for articles, but for adventures. It is my time and space, alone with my PC and my journal. I love reading about Victorian female adventurers and how they defied convention despite society's determination to shrink their horizons and their waists through second-rate roles and body-distorting garments. Women like intrepid explorer Isabella Bird (1831–1904), queen of the desert Gertrude Bell (1868–1926) who mapped out Iraq, and wayfaring biologist Mary Kingsley (1862–1900). They all left their inhibitions at home and journeyed into the unknown alone across Asia, Africa, Arabia and America. It took grit, especially in a man's world.

I decide I want to put female explorers 'back on the map', to rekindle a sense of adventure in myself and others, particularly my daughter Acadia and her co-millennials. I will revive a role model, an invincible woman who defied the status quo, walked on the wild side and explored the world without fear, eye liner or social media accounts. There are many; I cannot wait to compile a shortlist.

That is exactly what I am doing when I first meet Nellie Bly. Outside, a nippy day in winter 2013; inside, my desk edged up next to the radiator, scrolling, skimming and sifting through the lives of female adventurers. Suddenly she jumps off the screen at me. The more I get to know her, the more I am taken with this fearless woman who would not take no for an answer despite living at a time – the end of the nineteenth century –

when women 'knew their place'. Nellie knew her place all right, smack dab on the front pages of the world's newspapers. She sends all the other contenders scurrying. She pushes herself front and forward. I let her. We click.

I cannot resist Nellie's 'nothing is impossible' attitude that impelled her to conquer male-dominated newsrooms, feign madness to reveal brutality inside a women's insane asylum and whiz solo around the world with a single bag. Most of all, I am in awe of her humanitarianism. Nellie's newspaper campaigns for rights, justice and dignity gave voices to vulnerable people, especially women and children, and reformed corrupt institutions: asylums, prisons, sweat shops, orphanages.

Unlike Nellie, my undertaking to get female adventurers like her 'back on the map' will not be a race around the world. It will be an adventure. I will re-enact her journey. I will walk in her footsteps and I will witness her travels.

Nellie's first step was to study the routes and timetables of the ocean liners plying the seas to determine if she could circle the world in less than eighty days. Alas, the sun has set on the British Empire and I discover that the colonial ports that welcomed Nellie in 1889 now serve only cargo and cruise ships. The liners that Nellie knew are no more, and all of their routes, except for one, are lost at sea, along with my hopes of tracing her journey by ship. I must fly. I name my travel blog Nellie Bly in the Sky and begin to compile an itinerary.

The defining moment arrives for me when dates begin to fall in place. The year ahead will bring a milestone anniversary; 14 November 2014 marks 125 years since Nellie set off on her record-breaking voyage around the world. It is the sign I awaited; this landmark anniversary is the journalistic hook on which to hang my own journey. #NellieBly125 is born on Twitter; my blog goes live. I will re-enact Nellie's globetrotting journey to commemorate the 125th anniversary of her triumphant achievement.

*   *   *

When Nellie awoke the next morning, she knew exactly where to go – Steamship Row, where Manhattan meets Hudson Bay on lower

Broadway at Bowling Green. Here the offices of transatlantic liners stood shoulder to shoulder on cobbled streets scored with trolley tracks leading to the city's busy waterfront. Offering passages to ports around the world, Cunard Line, *Compagnie Generale Transatlantique*, Clyde Lines, Anchor Line and Red Star Line had set up headquarters in the Federalist-style former homes of city merchants. Their handsome brick three-storey façades with arched doorways and sash windows spoke of direct steamers between New York and Le Havre, Southampton, Antwerp and other destinations. Above their slate roofs capped with dormer windows and tall chimneys, steamship company names soared in giant letters visible far out into the harbour. It was one-stop shopping for Victorian voyagers.

The former Steamship Row site, today subsumed into New York City's throbbing financial heart, remains a magnet for travellers. Transatlantic ocean liner passengers like Nellie have been superseded by twenty-first century tourists charging onto ferries crossing to the Statue of Liberty and Ellis Island. These days only one authentic liner still plies the ocean and it is the largest, the longest, the tallest and the most expensive ever built. Cunard Line's flagship RMS *Queen Mary 2* crosses the Atlantic between New York and Southampton twenty-five times a year.

Glimpses of the golden age of steamship travel – when ocean liners ruled the waves and flying machines waited in the wings – linger in the Greek revival entrances for first class and cabin class passengers at Citibank, Number One Broadway. Mosaic shields representing major port cities, including London, Paris, Southampton and New York, ring Citibank's second storey like a necklace. Gargantuan ocean liners engraved in bronze decorate 20 Exchange Place nearby; and starred railings lead to the former offices of Blue Star Lines at 9–11 Broadway.

It was here, 125 years earlier, that Nellie calculated the crossings that could lead to the journey of her life, perhaps even the journey of the century. 'Anxiously I sat down and went over them,' she wrote, 'and if I had found the elixir of life I should not have felt better than I did when I conceived a hope that a tour of the world might be made in even less than eighty days.' Nellie was sure it could be done. She could travel the entire world in less than eighty days and break Phileas Fogg's record. With ships' timetables clutched in her hand and spinning in her head, she set off from Steamship Row to see her editor in *The World*'s offices at 31–32 Park Row on Printing House Square, about a mile away.

En route, Nellie's usual self-assurance vanished. Battling the cynical voices invading her head – too wild, too visionary, too far – she inched herself into managing editor John Cockerill's office that Monday. Without warning she blurted out: 'I want to go around in eighty days or less. I think I can beat Phileas Fogg's record. May I try it?' Cockerill looked up, a cautious smile emerging behind his walrus-style moustache as he toyed with the pens on his desk. He gave in eventually, but winning over *The World*'s business manager George W. Turner was another matter. The answer was an adamant no. 'The terrible verdict,' she called it. 'It is impossible for you to do it,' Turner said. Impossible was a word that Nellie abhorred. It held no place in her vocabulary and certainly not in her approach to life.

Turner laid out his reasons. 'In the first place you are a woman and would need a protector.' In the 1880s it was unthinkable for a single woman to travel alone around the city, let alone around the world, without a chaperone. Secondly, the sheer number of steamer trunks required would slow her down and impede her passage. And thirdly, Nellie spoke only English on an itinerary covering a dozen countries with at least ten languages. 'There is no use talking about it,' the business manager said. 'No one but a man can do this.' That was enough for Nellie. 'Very well,' she fumed. 'Start the man, and I'll start the same day for some other newspaper and beat him.' Turner reflected. 'I believe you would,' he said. By the end of the meeting Nellie had secured a promise that if anyone was commissioned to make the trip, it would be her.

Nellie had already established a reputation for doing whatever it took to get a good story, including putting her life in danger. Just the year before, when she had smuggled herself past the security guard at *The New York World*, Nellie was penniless, jobless and desperate. She had been pounding the city's pavements daily for four months to no avail. But before she left *The World* that day, Nellie had set the scene for an assignment that would transform her life, and the world of journalism, forever. Once inside the building, she finagled her way into the office of managing editor John Cockerill and proposed an article about the wretched conditions faced by destitute immigrants crossing the Atlantic to America. It was rejected.

But after conferring with *The World*'s publisher Joseph Pulitzer, Cockerill asked Nellie if she could work her way into an insane asylum. 'I don't know what I can do until I try,' she replied.

That was it; the conversation that ignited a brave new approach to journalism. This was the dawn of investigative journalism, and Nellie Bly was its pioneer. Just before leaving Cockerill's office she asked him how he would get her out of the asylum. 'I don't know,' was his bleak answer, 'only get in.' In her most astonishing achievement, even more than racing around the globe, Nellie convinced authorities that she was insane and endured ten excruciating days inside the women's section of the New York City Lunatic Asylum on Blackwell's Island. 'Positively demented. I consider it a hopeless case,' said one of the doctors who admitted her. His shocking verdict condemned Nellie, and countless other women, to ice cold baths in filthy water, putrid meals, relentless taunting from malicious nurses, and beatings that drew blood and bruises. 'What, excepting anguish, would produce insanity quicker than this treatment?' she wrote. Most unbearable of all for Nellie was the torment of her fellow inmates. 'In thinking of the greater misery of the others, I forgot the sting of my own.' Nellie found out how easy it was to enter an asylum, but now, to her increasing terror, she was discovering how hard it was to get out. This assignment from hell had landed her in a 'human rat-trap', a 'den of horror', that she feared she might never escape. At last Cockerill sent a lawyer to get her out after the 'ten longest days' of her life.

When it was published, *Behind Asylum Bars* shocked the nation. Nellie's accounts and her book *Ten Days in a Mad-House* unleashed sweeping reforms and almost $1 million to enact them. Inspectors invaded asylums across America seeking out cruelty and neglect. Her Blackwell's Island exposés brought acclaim for Nellie Bly. She had proven that she would go to any lengths for a story. But the assignment that would make her the most talked about woman on the planet was yet to come.

* * *

On the dark, drizzly evening of Monday, 11 November 1889, almost a year since her proposal to *The World*'s managing editor, Nellie received a note summoning her to his office. Heading downtown, her apprehension swelled into anxiety; what could she have done to warrant such an urgent summons? As she approached *The World*'s offices, a heroically sized bronze statue of Benjamin Franklin emerged from the ash-grey mist of Printing House Square. He carried a copy of *The Pennsylvania Gazette*, the newspaper he had published in Nellie's home state more than 100 years earlier. The effigy of this legendary printer, inventor, patriot and statesman, who once commanded the vast square flanked by City Hall and New York's leading newspapers, is now squeezed into an asphalt triangle. Today Benjamin Franklin rises from his massive granite plinth as testimony to the newspaper legacy of Park Row and Printing House Square, where Nellie carved out her career.

On this wet November evening, she was about to take on the assignment of her life. Cockerill was busy writing when Nellie slipped into his office. Without a word she sat down next to her editor's two-tier, multi-drawer desk and waited for him to look up. Manilla paper scrolls – marked, squashed and tossed – spilled from his wire waste basket. At last, lifting his neck and peering out under slumped eyebrows, Cockerill asked her quietly, 'Can you start around the world the day after tomorrow?' 'I can start this minute,' she replied. Less than seventy-two hours later, Nellie was aboard the SS *Augusta Victoria*. Her race was about to begin. There was just enough time to commission a travelling ensemble, purchase a warm coat and a travel bag, and write notes to her loved ones.

By mid-morning the next day Nellie was in the upper Manhattan studio of society dressmaker William Ghormley of Ghormley *Robes et Manteaux*, with shops in New York City and Paris. The four-storey brownstone townhouse, turned shop front, stood in the city's most prestigious commercial district frequented by the well-to-do 'carriage trade'. Nellie was known to take great pride in her wardrobe, so this may not have been her first visit to Ghormley's. 'I want a dress by this evening,' she said to Ghormley, 'a dress that will stand constant wear for three months.' William Ghormley did not even flinch. Bringing out a parade of fabrics, he draped his best cottons, linens, wools and silks artistically over a small table, studying their tones and textures in a large looking glass. He selected a plain blue broadcloth and a camel's hair plaid as the most durable combination for a travelling gown. Before Nellie left at lunchtime, the seams of her gown had been boned to give it structure, and she had already completed her first fitting. When she arrived back at 5.00 pm for a second fitting, the dress was finished. 'I considered this promptness and speed a good omen and quite in keeping with project,' she wrote. In eight hours Ghormley had fashioned a two-piece fitted gown that under normal circumstances would require at least four days of work. Nellie was delighted, but not surprised. 'If we want good work from others or wish to accomplish anything ourselves, it will never do to harbour a doubt as to the result of an enterprise,' she said. She need not have harboured any doubts. Ghormley acted as if it were an 'everyday thing for a young woman to order a gown on a few hours' notice'.

William Ghormley did not keep customers waiting. Earlier that year, he had created Caroline Harrison's gown for the inaugural ball celebrating her husband Benjamin's election as twenty-third President of the United States. The First Lady arrived at the ball on the evening of 4 March 1889 in a flowing fawn silk and brocade gown, to the music of John Philip Sousa and his Marine Corps Band. Ghormley's opulent design, complete with a train, is one of the most luxurious in the Smithsonian Institution's collection of First Ladies' ball gowns. Nellie's 'Ghormley gown' no longer exists and neither does Ghormley *Robes et Manteaux*. The celebrated society dressmaker did not stand the test of time; just four years after making gowns for the First Lady and Nellie, the business he opened in 1879 collapsed.[1]

The fitted blue broadcloth and camel hair gown delivered by Ghormley in twenty-four hours served Nellie non-stop for seventy-two days. More practical than eye-catching, her travel ensemble, including a Sherlock Holmes-style deerstalker cap, a black plaid ulster coat, and a small leather bag she called a gripsack, would soon be recognised the world over. 'I got a cap with a double-peak,' she wrote of her deerstalker. 'Quite English, you know!' Her iconic outfit turned up in newspapers and advertisements, on games and lookalike dolls.

* * *

Although her dress, coat and hat are lost to history, Nellie's small Swiss roll-shaped gripsack, known in Britain as a Gladstone bag, is the property of her biographer Brooke Kroeger. She loaned it for display at the

Newseum, a glass and steel superstructure in Washington DC devoted to journalism and freedom of expression, where I was lucky to view it before the institution closed in 2019. I had to see Nellie's gripsack. Just like feigning madness to get herself committed to a lunatic asylum, she would do whatever it took to race around the world faster than anyone ever had, even if it meant squeezing everything into a 16 x 7 inch bag, and wearing the same clothes for seventy-two days. Washington DC was not on my official Nellie Bly itinerary, but I quickly tacked it on and flew to meet my great friend and fellow journalist Louisa Peat O'Neil who had arranged a day at the Newseum. I was hoping that I could hold the gripsack – even just for a minute. But alas, we were separated by Plexiglas. The travel-worn leather bag, scuffed at the edges, was encased in a display applauding Nellie as an undercover reporter. Even though I knew its limited dimensions, and had seen photographs of Nellie carrying the gripsack, I was taken aback by its modest size, no bigger than a bolster cushion. Now I have my own, a vintage lookalike gripsack, exactly the same size, a gift from my French friend Delphine Higonnet, who snapped it up for me at a *brocante* in Belgium.

* * *

Nellie purchased her travel gripsack the afternoon before her departure. She chose her bag with the strict intention of confining her baggage. In her race against time, and the conventions that kept women 'in their place', Nellie had to be ready to move at a moment's notice. She also intended to prove that women were capable of travelling without trunks. Buying the gripsack was easy; but packing it was a nightmare: 'the most difficult undertaking of my life, there was so much to go into such a little space,' Nellie wrote. 'I got everything in at last except the extra dress. Then the question resolved itself into this: I must either add a parcel to my baggage or go around the world with one dress. I always hated parcels so I sacrificed the dress.' That's how Nellie came to circumnavigate the globe for seventy-two days in a single gown carrying one small bag.

At least thirty gripsacks the size of Nellie's would easily fit inside a single trunk. In an era when trunkless travel was unheard of, Nellie astonished porters, inspectors, ship stewards, fellow passengers, even the newspaper that employed her, with her lack of luggage. 'Just think of it – a run of 30,000 miles, more or less, and not even a Saratoga nor even a flat stateroom trunk,' *The World* stated, noting that 'many a belle' would feel hard done by without taking 'a round dozen of great roomy trunks for a fortnight's stay at a Summer resort'. In Nellie's era, massive curved iron-bound leather trunks known as Saratogas occupied most staterooms and ships' holds. Named after the swanky spa city and racecourse in New York, Saratogas featured separate compartments, pockets and trays for the ease of packing. *Hints for Lady Travellers*, a guide published the same year as Nellie's journey, warns women that gowns 'are the terrible part of packing; each one requires a tray to itself'. Nellie's only gown was on her back.

Dora de Blaquiére, writing for the *Girl's Own Paper 1890 Annual* on 'The Purchase of Outfits for India and the Colonies', praised Nellie as a shining exemplar of how to pack for long voyages, and noted that she was 'living proof' that fashion had at last 'gone hand in hand with common sense'. What Nellie didn't pack – crinolines, high heeled shoes, lace-up corsets and other body-deforming garments – impressed the *Girl's Own* journalist. 'While others debated on the best dresses for walking or working, the whole problem had been completely solved by this American lady in her two month and 11 day journey,' she wrote.

Even so, Nellie's gripsack bulged like a haggis. She squashed her belongings until everything fit, except for a gossamer silk waterproof that she draped across her right arm. Like closing the lid on a jack-in-the-box, she gave her contents a final shove and snapped the latch shut before it could spring open. 'One never knows the capacity of an ordinary hand-satchel until dire necessity compels the exercise of all one's ingenuity to reduce everything to the smallest possible compass,' she wrote.

Nellie's packing list:
two travelling caps
three veils
slippers
complete outfit of toilet articles
ink-stand
pens, pencils, and copy-paper

pins, needles and thread
dressing gown
tennis blazer
small flask and a drinking cup
several complete changes of underwear
liberal supply of handkerchiefs
fresh ruching fabrics to decorate collar and cuffs
jar of cold cream

'Most bulky and compromising of all' was the cold cream. She called it 'the bane of my existence'. 'It seemed to take up more room than everything else in the bag and was always getting into just the place that would keep me from closing the satchel,' she wrote. In Nellie's day, cosmetics were seen as instruments of the devil; only actresses and prostitutes used them. Respectable women applied cold cream, not makeup, to enhance their complexions; daubing it on their faces, hands and necks. Nellie used it to ward off the chapping that could come from travelling in foreign climates.

Along with her gripsack, Nellie carried £200 ($250) in English gold sovereigns, and Bank of England notes. She also took some American gold and paper money to discover if it would be accepted outside her country. The gold was tucked into her pocket; the notes were placed into a chamois-skin bag and tied around her neck. She kept track of local time with a timepiece attached to a leather bracelet; New York time was tracked on an ornate twenty-four-hour pocket watch.

What she *refused* to pack was a revolver. 'Someone suggested that a revolver would be a good companion piece for the passport, but I had such a strong belief in the world's greeting me as I greeted it, that I refused to arm myself.'

What she *forgot* to pack was a camera. 'The only regret of my trip, and one I can never cease to deplore, was that in my hasty departure I forgot to take a Kodak. On every ship and at every port I met others – and envied them – with Kodaks. They could photograph everything that pleased them; the light in those lands is excellent, and many were the pleasant mementos of their acquaintances and themselves they carried home on their plates,' she wrote. Alas, Nellie's round-the-world adventure is documented only by her words; she painted pictures with them. *The World* illustrated accounts of her journey with sketches: Nellie boarding the *Augusta Victoria*, eating breakfast on the train to Amiens, toasting with Jules Verne, the end of the journey at Jersey City, standing alongside history's greatest explorers. Many were drawn by her close friend at *The World*, cartoonist Walt McDougall.

Even with her ultra-lean luggage, Nellie believed that she could have packed even less. 'Experience showed me that I had taken too much rather than too little baggage,' she wrote. Leaving New York, she had no idea that whatever she needed, including ready-made dresses, could be purchased in ports along the way. Nellie packed 'on the theory that she would only be able to secure the services of a laundress once or twice' on her journey. In reality, laundry services were available at every port at prices far less than at home. Out at sea on the P&O ships, the quartermasters turned out a daily wash that would 'astonish the largest laundry in America,' she noted. Throughout her entire journey, only once did Nellie regret her limited travel wardrobe. Without an evening dress to wear, she chose to forego an official dinner in Hong Kong. But it was a small price to pay when compared with the 'responsibilities and worries I escaped by not having trunks and boxes to look after,' she wrote. 'So much for my preparations. It will be seen that if one is travelling simply for the sake of travelling and not for the purpose of impressing one's fellow passengers, the problem of baggage becomes a very simple one.'

*   *   *

Packing her bag was 'the most difficult undertaking' of Nellie's life. My biggest challenge is mapping my route. To organise her itinerary, Nellie had to make her way to the ocean liner offices on Steamship Row and leaf page by page through bundles of ships' timetables. All I have to do is switch on the Wi-Fi. I can plan it all online with an Around the World Explorer ticket. But I struggle as the planet stretches across the computer screen before me – continents and countries are powder-blue, oceans are sky-blue and airports are designated with navy blue dots. With so many places to take off and land, it looks like the map has measles. I wonder if everything is designated in shades of blue, the colour linked to serenity, to calm the nerves of people like me as we hop around the world on computer keys.

I am looking for a sign; any sign that will deliver me from the randomness of my route mapping. Nellie completed her journey in seventy-two days. My current itinerary fills twenty-seven days. Transposing the numbers is just the sign I need to hold my breath, close my eyes and press confirm on the Explorer ticket website. London–Colombo–Singapore–Hong Kong–Tokyo–New York–Washington–London, all direct flights, the shortest flying time available. My ticket confirmation arrives on 4 July 2014; accommodations are secured; the dream becomes a reality. I just need to purchase today's equivalent of Nellie Bly's leather gripsack.

The forty-litre black and grey rolling rucksack I buy for my journey is slightly larger than Nellie's bag. Measuring 21 x 10 inches, I can roll it on its two wheels or unleash hidden straps and carry it on my back. Billed as great for weekend adventures, it will serve me for more than a month as I follow Nellie around the world. Phileas Fogg carried a carpet bag, Nellie took a gripsack and I will travel with a convertible rucksack. I feel so righteous when travelling light. While others are organising porters as in earlier times, or swarming around baggage carousels like we do today, Nellie and I are off exploring our destinations. Along with independence and mobility, packing Nellie Bly-style means: no baggage fees, no lost or delayed luggage, easier access on public transport, and more space in hotel rooms or hostel dorms. It is also a plus when your accommodation is five flights up with no elevators.

Shrinking everything down to the 'smallest possible compass' like Nellie is easy with twenty-first century fabrics that wick, breathe, and even ward off rain, sunburn and mosquitoes. They pack down to nothing and dry overnight after a quick dip in the bathroom basin. My mix and match batch of clothing, in a quartet of basic colours: khaki, white, black and purple, takes up less space than Nellie's dressing gown alone, and gives me at least a dozen more outfits than her. My state-of-the-art tablet computer, weighing slightly over two pounds, replaces Nellie's ink-stand, pens, pencils and copy-paper, although I am never without a reporter's pad and biro. Essential communications gear, not cold cream, clogs my rolling rucksack: the tablet, camera, mobile phone, chargers and their accompanying adapters. The only lipstick I take will get lost along the way.

My packing list:
two skirts (one khaki, one floral)
a navy mosquito-resistant shift
khaki shorts
three t-shirts: white, purple and lime green
one gingham shirt
chinos
black pants
rain jacket*
swimsuit
several complete changes of underwear*
sewing kit*
toiletries*
first aid kit
glasses and sunglasses
tablet, mobile phone, camera, chargers and adapter
notebook* and pens*
running shoes
sandals
flat shoes
flip flops
four bandanas and a shawl
microfibre towel
mini daypack

* the same as Nellie

My bandanas will multi-task as scarves, serviettes, washcloths, hankies and mini tablecloths. Purple, pink, turquoise and red, I can wear them around my neck, in my hair, on my head and in my pocket. The shift triples as a sun dress, beach cover-up and nightgown. Protection from too much sun or chilly winds, my black and white striped cotton shawl from India adds a dash to travel outfits, covers my legs in temples, and provides something to sit or lie on. I wear it over my head, across my shoulders and around my waist. I secure my case with a mini pale blue combination lock, more as a deterrent than proper protection. My secret code is 125, marking the 125th anniversary of Nellie's voyage. Nellie did not lock her gripsack; it was too full.

*   *   *

Nellie wrote letters to her loved ones the night before she left. My loved one, my daughter Acadia, 19, wrote a letter to me.

Dear Mummy

You're leaving in five days! Although I'm a little apprehensive about the trip and wish you were travelling first class, I am very proud that you are going round the world when most mummies are grey and boring and watching *The Morning Show*. So I guess you do classify as a 'cool mummy.'

I am so jealous, please can we go travelling my style, five-star hotels, spas, first class, when I've finished university?

On your travels, remember be interested, not interesting. Unless it's in a sealed bottle, don't drink it and if it's not from a pack, don't smoke it.

You don't really tell me off unless it's about the important things. You've always supported me, no matter how many silly and occasionally slightly immoral things I've done and you always seem to be right in the end!

I'll try not to stress Daddy out too much and manage my money a bit better while you are away. Also I'm being taught to cook by my roommates so you might have competition when you get back; how I'll survive without your shrimp risotto for a month I have no idea.

I love you lots and lots.

Wormie [My nickname for Acadia because she wiggles a lot and loves to read books.]

Chapter 2

In Which Nellie Crosses the Atlantic

Navigation sculpture at former P&O offices, London

Hoboken, New Jersey; Southampton and London, England
14–22 November 1889

*'But when the whistle blew and they were on the pier, and I was on
the* Augusta Victoria, *which was slowly but surely moving away from
everything I knew, taking me to strange lands and strange people, I felt
lost.'*

Nellie Bly

Nellie was not an early riser. That day she had no choice but to 'get up with the milkman'. The journey of her life was about to begin. She awoke early, turned over a few times, dozed off and woke up again with a start 'wondering anxiously if there was still time to catch the ship.' She scolded 'the good people who spend so much time in trying to invent flying machines' saying they 'should devote more energy to promoting a system in which boats and trains would always make their start at noon or afterwards.' This was not a good precursor for the unprecedented global journey she was about to undertake. In the end, the delays Nellie endured were not her fault; they can only be blamed on weather and the steam-powered transport delivering her from port to port, station to station around the world. Nellie was powered by determination. 'I always have a comfortable feeling that nothing is impossible if one applies a certain amount of energy in the right direction. If you want to do it, you can do it,' was the motto she lived by.

She put off her departure from home for as long as she could. 'There was a hasty kiss for the dear ones, and a blind rush downstairs trying to overcome the hard lump in my throat that threatened to make me regret the journey that lay before me,' Nellie wrote. Before she could change her mind, she jumped aboard a streetcar to the Christopher Street depot for the ferry to Hoboken. Nellie made her way alone that morning, accompanied only by sporadic pangs of anxiety. She tried her best to ignore them, but she was haunted by the impending perils she could face – bitter cold, intense heat, raging storms, shipwrecks, disease, loneliness and worst of all, the humiliation of losing her race against time. She knew that winter, when typhoons and monsoons ravage the oceans, and blizzards block the railways, was the worst possible time to be leaving. As she approached the docks in Hoboken, Nellie realised the magnitude of the journey that lay before her.

Some friends who knew of her hurried departure were at the docks to wave Nellie off. 'The morning was bright and beautiful, and everything seemed very pleasant while the boat was still; but when they [her friends] were warned to go ashore, I began to realize what it meant for me,' Nellie wrote. '"Keep up your courage," they said to me while they gave my hand the farewell clasp.'

Nellie tried to be brave. 'But when the whistle blew and they were on the pier, and I was on the *Augusta Victoria*, which was slowly but surely moving away from everything I knew, taking me to strange lands and strange people, I felt lost,' she wrote. 'My head felt dizzy and my heart felt as though it would burst … the world lost its roundness and seemed a long distance with no end, and – well, I never turn back.' She gazed at the pier, watching her friends until she could no longer see them as the city she was leaving dissolved into shadows. 'I am off,' she said to herself from the deck of the *Augusta Victoria*, 'and shall I ever get back?'

Nellie had never travelled by sea before. When she stepped onto a ship for the first time in her life, she did not get off again for eight turbulent days. Land had always been within sight, now all Nellie could see was an endless, angry Atlantic that seized the SS *Augusta Victoria* soon after she left the harbour and rarely let loose. Seasickness took its toll, but only for one stomach-churning day. 'Never having taken a sea voyage before, I could expect nothing else than a lively tussle with the disease of the wave,' she wrote. What really twisted Nellie's stomach in knots, far more than seasickness, was the prospect of delays from the brutal headwinds threatening her transatlantic voyage.

Inside the Hamburg America Line's first luxury liner, Nellie spent the days at sea dining beneath ceilings painted like the sky, taking tea among potted palms in the winter garden, and relaxing on button-backed leather settees in the ladies' lounge. The SS *Augusta Victoria* was a floating hotel, its grand rooms furnished with gleaming mahogany, domed ceilings, stained glass and sweeping staircases with filigree balustrades. Gilded cherubim graced mirrors and lamps.

On deck, waves lashed the railings, sending wooden lounge chairs gliding like skates. The ropes securing the liner's three masts trembled and groaned in the winds. Six months earlier, on her maiden voyage, the SS *Augusta Victoria* had become the fastest liner on the Atlantic. Now she was competing against mighty squalls to maintain her record, and deliver 300 bags of mail and 1,100 passengers to Southampton by 10.00 am on 21 November. At lunchtime that day, after seven days at sea, Nellie glimpsed land for the first time since leaving home. To her the rock-strewn Lizard Peninsula, Britain's southernmost point, seemed 'the most beautiful bit of scenery in the world'.

Southampton was asleep when the SS *Augusta Victoria* steamed into port the next morning at 2.30 am, sixteen hours late. The tugboat to take Nellie ashore arrived thirty minutes later with Tracy Greaves, the London correspondent of *The New York World*, on board. He would be her escort in England and France. The end of her transatlantic voyage signalled the beginning of Nellie's journey across Europe. The first leg of her travels was now complete. 'On the Other Side', *The New York World* shouted from the front page of its 22 November 1889 edition after receiving a cable dispatch from Nellie. 'She has passed the first milestone on the roadway to success.'

Another milestone was in the making as the tug cast off from the SS *Augusta Victoria* and drifted into the dark. 'Mr and Mrs Jules Verne have sent a special letter asking that if possible you will stop to see them,'

Greaves told Nellie. The prospect of meeting the author of *Around the World in Eighty Days*, the book that inspired her journey, thrilled Nellie. 'Oh how I should like to see them,' she said. But the visit would require a time-guzzling deviation that could threaten her quest right from the start. Nellie was only eight days into her race to beat the record set by Verne's character Phileas Fogg. Her schedule was tighter than a ship's rigging, but how could she resist an invitation to visit Jules and Honorine Verne? It could be done, Greaves told her, if she was willing to sacrifice two nights of sleep, travel 180 miles off-route, insert two extra train journeys and add fourteen hours to her intense itinerary. 'Safely? Without making me miss any connections?' she asked Greaves. Once reassured by the London correspondent, Nellie dismissed the need for sleep and accepted with pleasure the invitation to the Vernes' home in Amiens, northern France. There was a hitch, however; they must get to London before sunrise. The last train from Southampton had already gone; the next scheduled departure would be too late to make the vital connections. But if, and only if, an extra train was organised to deliver the mail bags delayed by the ship's late arrival, visiting the Vernes should be possible.

They were in luck. As the mail bags were being loaded, the two journalists joined a handful of passengers on a special mail train to London's Waterloo station. Settled in maroon leather seats, they attempted to ward off the pre-dawn chill with rugs and iron foot-warmers provided by the porter. An oil lamp emitted more smoke than light, flooding the railcar with a stench that stung their eyes during the two-hour journey through southern England. They were 'flying over the rails to London,' Greaves said. When they arrived at 5.00 am, Waterloo station was as deserted as a football stadium on a non-match day. Nellie and Greaves had just four hours to obtain an official passport from the American Legation, visit *The New York World*'s London office, go to the Peninsula and Oriental Steam Navigation Company (P&O) to purchase tickets for the remaining crossings, and catch the express train to Folkestone for the ferry to Boulogne. They did it, just. Emerging from Waterloo station they were met by 'as gloomy and foggy a morning as this gloomy and foggy old city ever saw,' according to Greaves. To Nellie, the fog 'hung like a ghostly pall over the city.' They hailed the only cab around that early, a four-wheeled horse-drawn Brougham. Nellie peered out from the carriage as Greaves pointed out the River Thames, Westminster Abbey and the Houses of Parliament. Piercing the mist, the spired turrets of Parliament gave her a sense of the fine Gothic Revival buildings beneath. Nellie was fond of fog; it reminded her of Pittsburgh where she lived before moving to New York. She liked how it 'lends such a soft, beautifying light to things that otherwise in the broad glare of day would be rude and commonplace.'

The pair navigated Victorian London along cobbled streets and broad stone pavements. I track their journey below the surface on the London Underground, on buses and on foot. London is where I live and the 'headquarters' for my expedition to replicate Nellie's record-breaking voyage 125 years after she set off. In two weeks' time I will begin my travels to far-off, unfamiliar places. Following her in my hometown is a cinch, but picturing her dashing through this twenty-first century mega-city is not. It is a cloudless Sunday afternoon in August. I try to envision the chilling November fog penetrating Nellie's thick wool ulster, bringing on the shivers after a night robbed of sleep.

The first stop on the pair's 'mad scamper across London' was the *New York World* office on Cockspur Street near Trafalgar Square where Nellie picked up her cables and instructions on where to obtain her official American passport. In the urgency to begin the race, Nellie had left America with only a temporary special passport, number 247, personally organised by no less than US Secretary of State James G. Blaine. From Cockspur Street, the pair were sent to the West End residence of the Second Secretary of the American Legation, Robert Sanderson McCormick. An ambitious young diplomat with a waxed moustache that twisted up towards his ears, Second Secretary McCormick had risen early to complete the paperwork that would open doors to Nellie around the world. McCormick was representing the US Minister to the Court of St James, Robert Todd Lincoln, son of the late President Abraham Lincoln. The Legation was based near Victoria station in 'small dingy offices' at 123 Victoria Street. Lincoln is said to have found them so objectionable that he rarely hosted delegations there, according to US Embassy sources.

Nellie had been travelling all night; she was exhausted and hungry. But that was no excuse for her to bend the truth when answering official questions posed by McCormick. To finalise her passport, Nellie had to swear to the date of her birth. Knowing it could be a sensitive question, 'the one question all women dreaded to answer,' the Secretary asked Greaves to distance himself. Nellie was having none of it. 'I will tell you my age, swear to it too, and I am not afraid; my companion may come out of the corner,' Nellie said. Despite her apparent bravado, it later emerged that Nellie took this opportunity to extend her youth. She stated that her birthday was 5 May 1867; instead of 1864.[1] She made herself 22 when she was 25. It was a very Victorian thing to do, a very Nellie thing to do. This little white lie causes confusion to this day as books, blogs and even commemorative plaques muddle the two dates.

*   *   *

If they exist, I cannot find any architectural traces of *The World*'s offices on Cockspur Street and I do not have an address for the Second Secretary's West End residence, but I can find the footprint of the former American Legation snubbed by the President's son at 123 Victoria Street. It is subsumed beneath eleven floors of swanky glass-fronted offices, now headquarters to corporates like the John Lewis Partnership and luxury brand Jimmy Choo. Yesterday's modest legation has transformed into today's $1 billion American Embassy, a crystalline fortress set just off the River Thames in southwest London, a testimony to America's rapid ascent to world power. Only 125 years earlier, Nellie's travels revealed that the 'majority of foreigners' could not pinpoint America on a map. Indeed, Nellie discovered that the only country that recognised US currency was Ceylon, but at a prohibitive exchange rate. She relied entirely on her Bank of England notes and coins.

Nellie and Greaves will have encountered the Bank of England on their way to the P&O Steamship Company on Leadenhall Street where they purchased passages from Brindisi, Italy, around the world to San Francisco. The shipping company's 1840 headquarters at 122 have been replaced by the sleek, chic and shiny Leadenhall Building hurtling forty-eight storeys into the sky. This wedge-shaped landmark, designed by the celebrated Richard Rogers Partnership, is known fondly by Londoners as the Cheesegrater. Below the streamlined steel and glass, in the Cheesegrater's undercroft, the architects have retained a poignant reminder of the bygone age of steamship travel in a carved wall relief entitled *Navigation*. A bearded, God-like figure cradles an ocean liner in his right hand and grasps a ship's wheel in his left. His body-builder physique is scantily clad in stone fabric discreetly draped to conceal his maleness. An Odysseus lookalike, he could be the Greek hero of Homer's poems who wandered the seas for ten long years on his way home from the Trojan War.

The final destination on Nellie's four-hour dash across London was Charing Cross railway station, opened in 1864, a few months before she was born. There was just enough time for the journalists to swallow a few mouthfuls of ham and eggs, and knock back some coffee at the Charing Cross Hotel before the 9.00 am train to Folkestone was announced. 'I know that cup of coffee saved me from a headache that day,' Nellie wrote. 'I had been shaking with the cold as we made our hurried drive through London, and my head was so dizzy at times that I hardly knew whether the earth had a chill or my brains were attending a ball.'

After a morning of traipsing after Nellie and Greaves, I treat myself to an over-priced cappuccino in the hotel's Terrace bar overlooking the Strand. It arrives with a delicate white heart painted in froth. In the spirit of the past, I have travelled here on a classic double-decker Routemaster bus dating back to the 1950s. The old Routemasters, like some of the sites I have tracked today, were discarded for twenty-first century replacements, with one exception. The Number 15, travelling from the City to the West End, is now classified by Transport for London as 15H. H stands for heritage, which is in abundance here at the Charing Cross Hotel, a Grade II listed building. Nellie, Greaves and I have entered the same place, right down to the French Renaissance-style façade, the six surviving tracks and the station clock in crimson, gold and black. The very same clock was keeping time when the final announcement sounded for the train to Folkestone. Nellie and Greaves made it on board by a whisper. As the train steamed out of the station, the fog that had masked London all morning was lifting. From her carriage Nellie could see the gold-laced clock faces of Big Ben and the great dome of St Paul's Cathedral, her final glimpses of London. She was on her way to France to visit Jules Verne in hopes that she would be the first person to turn his fiction into reality.

On 2 October 1872, Verne's fictional hero and his French valet Passpartout departed from Charing Cross on the 8.45 pm train for Dover. Since then, Phileas Fogg's daring global quest has ignited a parade of globetrotters inspired by the ultimate journey. Nellie was the first. She left London from the same station in 1889. Ninety-nine years later, Michael Palin set off from Victoria station to replicate Phileas Fogg's adventure for the BBC travelogue *Around the World in 80 Days*. Michael Palin followed Verne's globetrotter; I am following mine.

Chapter 3

In Which Nellie Meets Jules Verne

Jules Verne's home, Amiens

London, England and Amiens, France
22 November 1889

*'I had gone without sleep and rest; I have travelled many miles out of
my way for the privilege of meeting M and Mme Verne, and I felt that
if I had gone around the world for that pleasure, I should not have
considered the price too high.'*

Nellie Bly

A sleep-deprived Nellie opted to snooze on the train from London to Folkestone. Through the window, fields and farms stippled in autumn mist unrolled across the downlands that eventually tumbled into the Channel. But the arcadian landscapes of Kent, admired then and now as 'the garden of England', were lost on Nellie. 'What is scenery compared with sleep when one has not seen bed for over twenty-four hours?' she said to Greaves before nodding off. Images of her mother, her friends and her neighbourhood drifted through Nellie's dreams that morning, transporting the globetrotter back to New York City, now more than 3,000 miles away. She awoke as the train squealed into the Folkestone Harbour railway station, catapulting her straight back to matters at hand. Here the second ship she had ever boarded awaited

Nellie at the wharf. This time she would be crossing the English Channel to the port of Boulogne-sur-Mer in France, the third of a dozen countries on her itinerary.

The two journalists could have been travelling under the sea from Folkestone, instead of across it, if work on an early tunnel had not been abandoned seven years earlier in 1882. Machines burrowing through marlstone on both sides of the Channel were halted amid fears that a link to mainland Europe would undermine British defence and put the nation at risk.[1] Today's Eurotunnel, the biggest civil engineering project of the twentieth century, opened in 1994, almost 200 years after it was first proposed. In Folkestone, a maze of rails, ramps, bridges and tunnels swirl across 370 acres, sending 60,000 passengers a day through the longest undersea passage in the world, England to France in thirty-five minutes. A nineteenth-century tunnel would have whizzed Nellie under the Channel and soothed her jitters about traversing what she called the 'stream of horrors' feared even by 'hardy sailors'.

Greaves described the 'stream of horrors' that afternoon in late November as a 'choppy little channel in a cross mood'. Waves flailed as if they were drunk, thrashing the ship as it ploughed thirty-two miles to the French coast. A pair of black and white funnels belched grey smoke trailing the steamer like superhero capes. The musty, airless cabin below sent Nellie above board despite the chill winds that swept the deck and rocked the ship. Greaves said Nellie 'sniffed at the obvious discomforts of a Channel crossing' and remained on deck because she was already 'inured to the terrors of the sea' by her turbulent transatlantic passage. Nellie's nose was as cold as the violent waters below as she gazed at the seagulls wheeling overhead. Hissing from blazing steam boilers riddled the air as bitter winds stung her cheeks. With Folkestone and the British coast vanishing behind her, Nellie took stock of her journey so far. 'So I have been to England, have I?' she said to a shivering Greaves. 'Well I like it, what I have seen of it, and I am coming again.' After ninety minutes at sea, she could almost distinguish the wedding cake dome of the Basilica of Notre-Dame de Boulogne rising in the hills above the port. Before long they were approaching the Quay de Pacqueboats on the Liane River, the main waterway from the Channel to the bustling port, where steamships and fishing boats shared the same docks. Today only the fishing boats remain; the last Channel crossing here was in 2010. More than 150 trawlers painted in primary colours and draped in tangerine fishing nets anchor side by side where steamers once docked. Boulogne-sur-Mer is now France's leading fishing port.

Nellie and Greaves set off for Amiens from the Gare de Boulogne-Maritime, commanding the pier where steamships linked with the railway. As their train veered from the coastline into the countryside, low clouds hung like grey blankets over naked trees and fields gleaned of their harvests. 'A touch of winter,' wrote Nellie. Two hours later, they were approaching the Gare du Nord, now the Gare d'Amiens. On the platform awaiting them were Jules and Honorine Verne with *The World*'s Paris correspondent Robert Harborough Sherard, sent to act as translator. Sherard was an Oxford-educated English author, confidant of Oscar Wilde, and the great-grandson of poet William Wordsworth. He considered this assignment 'somewhat foolish, but amusing'.[2] Regarding Nellie through *pince-nez* spectacles, he only ever referred to her as 'the girl'. But like Nellie, he could not resist the prospect of meeting the world-renowned French writer.

Jules Verne clasped a spray of fresh flowers to present to his American guest. Stepping from the train carriage into the station's soaring cast iron and glass hall, Nellie fretted about her appearance. 'I wondered if my face was travel-stained, and if my hair was tossed,' she wrote. A train in her own country would have afforded Nellie the privacy to make her 'toilet *en route*' and be as 'trim and tidy' as if she were receiving the Vernes in her own home. But there was no time for regrets. 'They were advancing towards us, and in another second I had forgotten my untidiness in the cordial welcome they gave me. Jules Verne's bright eyes beamed on me with interest and kindliness,' she wrote, 'and Mme. Verne greeted me with the cordiality of a cherished friend.' The formalities 'that freeze the kindness in all our hearts' were dispensed with here. In minutes Jules and Honorine Verne had won Nellie's 'everlasting respect and devotion'.

Climbing into two horse-drawn carriages parked in the forecourt, the party departed for the Vernes' residence in the newly-built *bourgeois* district of Henriville. As they trotted through the falling twilight, Nellie glimpsed 'bright shops, a pretty park and numerous nurse maids pushing baby carriages about'. In twenty minutes they arrived in the paved courtyard of the Vernes' red brick *hôtel particulier*, a grand townhouse distinguished by its imposing five-storey circular tower. Greaves described the Vernes' home as 'the finest in Amiens, standing in extensive grounds, with a superb view of the cathedral and the city'. The couple's large, black and shaggy dog Follet jumped on Nellie in a 'loving welcome' that she did not reciprocate for fear of losing her dignity at 'the very threshold of the home of the famous Frenchman'.

Almost 125 years later, we are on our way to Gare d'Amiens, same place, new station. On the day before Easter, my husband David, daughter Acadia, 19, and I are re-living Nellie's time with the Vernes. The official inauguration of my Nellie Bly quest, it is the first of many journeys I will take in her name, and the only one with all three of us. It feels right, even auspicious, to begin here in the northern French city of Amiens where Jules Verne wrote *Around the World in Eighty Days*, the book that triggered Nellie's journey, and therefore my own. My next Nellie Bly-inspired journey will lead me much farther afield to the South Asian island of Sri Lanka, Ceylon in Nellie's day. That trip is in September, five months down the line.

The cathedral city of Amiens in Picardy is an hour northeast by car from the mediaeval site of Gerberoy, where we spend many of our holidays. We have owned a house in Gerberoy since 1999 when David was posted to UNESCO headquarters in Paris to represent Great Britain. One of France's 160 *Plus Beaux Villages*, Gerberoy lies about half-way between Paris and Amiens. We know this region well and have often visited Amiens to admire its glorious Gothic cathedral, a UNESCO World Heritage Site. But we have never seen the city through Nellie's eyes, or even Jules Verne's. Now we will.

Early spring sunlight illuminates the leaves beginning to unfurl on many of the trees as we follow the D930 towards Amiens. A cloud of swallows returning from Africa coasts so low we can see their ruby throats and forked tail feathers. Before long the countryside surrenders to the city as we follow signs to the *centre ville*. The cathedral's two towers and heaven-high octagonal spire, the largest in France, rise in the distance. Pulling into the Gare d'Amiens, we scamper out of our 2006 Volvo saloon and head for the platforms to pretend we have arrived by train like Nellie and Greaves. Judging by the clever deployment of glass and cement, we are not surprised to learn today's station was designed by architect Auguste Perret, master of reinforced concrete. A station entrance is named for him. Far more streamlined than its ornate *Beaux-Arts* predecessor that Nellie knew, the post-war station still serves Boulogne, Calais and Paris. An original 1955 railway clock states the time in roman numerals; it is almost XI o'clock. A regional *Nord Pas de Calais* express train, branded in bright blue and yellow, is preparing to depart for Calais on Nellie's route. Engines hum and warning beeps sound as the carriage doors shut automatically. The train glides out of Gare d'Amiens heading north along the coast as it did when Nellie and Greaves were onboard. We rejoin the Volvo and motor over to the Maison de la Tour, Jules Verne's former home on the Boulevard de Longueville, now Boulevard Jules Verne.

In his day, the author could see the cathedral and the train station from his window; he could hear the trains too. Railway lines, replacing the city's twelfth-century ramparts, run below the embankment outside his house. The clattering rails, whistles and bursting steam reminded Verne of city life in Paris, his home before moving to Amiens in 1871 at the request of Honorine. In 1882, they settled in the Maison de la Tour, where Nellie paid her celebrated visit. Now known as Maison Jules Verne, visitors from around the world come here to pay homage to the father of sci-fi and the world's most translated writer after Agatha Christie. Maison Jules Verne is the French equivalent of Beatrix Potter's Hill Top in Cumbria and Shakespeare's Birthplace in Stratford-upon-Avon, two of my favourites.

David, Acadia and I enter the panelled carriage gates that lead into the courtyard where Nellie, seated next to Mme Verne, drew up in a carriage. Acadia, dressed in blue jeans and black suede heels, poses on the steps of the maison's canopied front entrance. Five marble steps lead inside to the winter garden where amethyst orchids bloom and lemons dangle from miniature trees stationed on a vast marble floor chequered in black and white. Outside, in the real world, it is a spring-time Saturday; inside it is 22 November 1889 and we are with Nellie.

Her visit lives on here. Nellie's account of the Vernes' *grand salon* in her own book *Around the World in Seventy-Two Days* is framed and hung here for all to read:

The room was large and the hangings and paintings and soft velvet rug, which left visible but a border of polished wood, were richly dark. All the chairs artistically upholstered in brocaded silks, were luxuriously easy. Beginning at either side of the mantel they were placed in a semi-circle around the fire, which was only broken by a little table that held several tall candlesticks.

When Nellie was here, the *grand salon* was 'dusky with the early shade of a wintry evening'. A fire crackled in the grate, 'casting a soft warm light' on the party of five. Once everyone was seated, a fine white Angora cat wandered in. Nellie felt its silky fur on her knee before the cat climbed into Mme Verne's lap to be stroked. Jules Verne once said he believed cats to be 'spirits come to earth'. 'A cat, I am sure, could walk on a cloud without coming through,' he said. Nellie was seated next to the mantelpiece in an armchair closest to the open fireplace. Sherard sat on her right and translated for the others in 'an attractive, lazy voice,' according to Nellie. The Paris correspondent later described his host as having 'much in his face that reminds one of Victor Hugo' and compared him to a 'fine, old sea captain, ruddy of face and full of life'. Nellie could feel Verne's sheer love of life as he balanced on the edge of his chair to address her. 'The rapidity of his speech and the quick movements of his firm white hands bespoke energy – life, with enthusiasm,' she wrote. She loved the way his long and thick white hair stood up in 'artistic disorder' and the 'brilliancy of his bright eyes' that were overshadowed by snowy brows. But in Nellie's view, Honorine Verne, seated next to London correspondent Greaves, was by far 'the most charming figure in that group'. Mme Verne's eyes were bewitchingly black; her smiling lips were rosy red, Nellie wrote. She was dressed in a watered-silk skirt and a bodice of black silk velvet very becoming to her short, well-rounded figure. 'Imagine a youthful face with a spotless complexion, crowned with the whitest hair, dressed in smooth soft folds on the top of a dainty head that is most beautifully posed on a pair of plump shoulders,' wrote Nellie, 'and you have but a faint picture of the beauty of Mme Verne.'

Jules Verne was incredulous that 'the slight young woman could possibly be going around the world all by herself'. 'Why she looks like a mere child,' he told Sherard. Verne was taken with Nellie's readiness to jeopardise her race to visit him in Amiens. 'Her coming out all the way to see me was a risky undertaking,' said the author. 'I feel as much interested in the success of this enterprise as I did when I was following Phileas Fogg and his friends round the world in my study.' He said to Greaves that '*The New York World* must be a wonderful paper if it has many like Nellie Bly on its staff'. 'No,' replied the London correspondent, 'there is only one Nellie Bly.'

We stand now, David, Acadia and I, where they all were once sitting. The fireplace is empty, and the cat is but a ghost, but the soft velvet rug remains, as do the dark chairs upholstered in brocaded silks. The candle-laden crystal chandelier, garlanded wall mouldings and Louis Quinze mirror above the mantel whisper of *Belle Epoque* France. We seek Nellie's presence in this lavish salon, also called the music room, where Honorine Verne loved to host weekly parties.

For the very first time, Nellie and I are in the same room. I have a feeling there will not be many moments like this when I can be so sure. Here at the Maison Jules Verne, we are in an enclosed space; our encounter is inevitable. Her footsteps will grow fainter on city streets, along esplanades, at ports and in train stations where I may not be able to find her, to distinguish her. Many of the places that defined Nellie's voyage will have vanished over time. The ocean liners are no more; few of the hotels she passed through have survived. My adventure is about seeing what she saw and doing what she did. Although I will follow in Nellie's footsteps with the nose of a bloodhound and the skills of a journalist, I know that I may never be so close to her again. Her photograph, the classic head and shoulders portrait with her hair pinned up over a sweeping brown fringe, a crocheted starburst collar embellishing her dark silk embossed dress, is displayed in a curio cabinet here. It was a souvenir of her visit.

As a reporter like Nellie, I like to think I would have asked Jules Verne similar questions:

'Have you ever been to America?' Answer: 'Yes once, for a few days only, during which I saw Niagara. I have always longed to return but the state of my health prevents me taking any long journeys,' he replied. Verne was referring to a severe leg injury that left him with a limp. 'I know of nothing I long to do more than to see your land from New York to San Francisco,' he told Nellie. 'I try to keep a knowledge of everything that is going on in America and greatly appreciate the hundreds of letters I receive yearly from Americans who read my books.' Verne had a collection of over 2,000 letters from America.

'How did you get the idea for your novel?' Answer: 'I got it from a newspaper.' It was in an article in the French newspaper *Le Siecle* showing calculations on travelling around the world in eighty days that Jules Verne discovered the basis of his novel. The article had not taken into account the difference in the meridians which gained a day for Phileas Fogg and meant he won his bet. Had it not been for what he called 'this *denouement*', Jules Verne told Nellie he would never have written *Around the World in Eighty Days*.

There could have been another inspiration that Verne didn't share with Nellie, the basis for this and other visionary voyages born on the pages of his novels like *Twenty Thousand Leagues Under the Sea* and *A Journey to the Centre of the Earth*. Legend has it that when Jules was 11, he secretly secured a job as a cabin boy on a ship travelling to the Indies. Before it set sail, his father intercepted the ship and removed his son. Pierre Verne did not scold Jules, but he exacted a promise from his son that he would 'travel only in his imagination'. The rest is history.

Jules Verne travelled by 'imagination', by land and especially by sea in a boat named after his son, *Le Saint-Michel*. Through his writing, he encouraged others to do the same, particularly youth: 'young people, travel if you can, and if you cannot – travel all the same!' 'Travel,' he said, 'enables us to enrich our lives with new experiences, to enjoy and to be educated, to learn respect for foreign cultures, to establish friendships, and above all to contribute to international cooperation and peace throughout the world.'

During the visit to the Maison de la Tour, Jules Verne told Tracy Greaves that his greatest literary endeavour was to describe the Earth by means of novels. 'I would like to leave in my books pictures of the whole of the globe,' he said. In the most celebrated outcome, *Around the World in Eighty Days*, Verne's best-loved character Phileas Fogg is said to be based on his father Pierre, who was relentless in his time-keeping.

Time was of the essence for Nellie. Casting an eye on the filigree Louis Quinze clock that hung on the wall opposite her in the salon, Nellie saw the moments slipping away. There was only one train she could take from

Amiens to Calais. If she missed it she would be delayed for a week and she might as well give up the race. All the same, she wanted to visit her host's study before their departure.

Candlestick in hand, Mme Verne guided the way, lighting gas lamps as they ascended a spiral staircase to her husband's private domain. A haze of gaslight and candlelight cast shadows along the narrow passage leading to the little corner room that served as the author's study. A simple camp bedstead stood against the wall, a small table was pulled up next to the window. Nellie was astonished. 'When I stood in M Verne's study, I was speechless with surprise. The room was very small; even my own little den at home was almost as large. It was also very modest and bare. One bottle of ink and one penholder was all that shared the desk with the manuscript.' The manuscript was said to be his latest novel, *Sans Dessus Dessous* (*Topsy-Turvey or The Purchase of the North Pole*), about Americans who, for the sake of speculation, make an attempt to change the axis of the Earth to convert polar regions into a fertile garden. The 'extreme tidiness' of Verne's manuscript impressed Nellie, no additions scratched in the margins, only deletions on the page. It made her think that 'Mr Verne always improved his work by taking out superfluous things and never by adding.' Verne's discipline, as recorded by Nellie, is sterling guidance for all writers, and I try to follow it.

The author's study, at the heart of the household, is the smallest and most intimate room in an otherwise opulent residence. It is where visitors feel closest to him. David and I have it to ourselves; Acadia has slipped off to the attic where models of the extraordinary flying machines Verne invented – the *Albatross*, the *Go Ahead* and the *Épouvante* – hang from the ceiling. Almost everything in his study is green: moss wallpaper swathed in ivy, floor-length curtains secured with braided olive tiebacks, even the paintings depict trees and forests. The pedestal globe is the author's own. The world has certainly faded, and the geography has definitely shifted. More than thirty novels came to life in this study where Jules Verne worked from 5.00 to 11.00 am every day. 'Here in this room with these meagre surroundings, Jules Verne has written the books that have brought him everlasting fame,' wrote Nellie.

The time had come to depart. Back in the *grand salon*, before the blazing fire, they shared a glass of wine and a biscuit before bidding each other farewell. 'In compliment to me,' Nellie wrote of her host, 'he endeavoured to speak to me in English, and did succeed in saying, as his glass tipped mine: 'Good luck, Nellie Bly.' Mme Verne was not going to be outdone by her gallant husband. The author's wife told the Paris correspondent that she would like to kiss her guest goodbye, a great tribute to Nellie. Accepting a gentle kiss on each cheek from Mme Verne, Nellie 'stifled a strong inclination to kiss her on the lips … and show her how we do it in America.' 'My mischievousness often plays havoc with my dignity,' she wrote, 'but for once I was able to restrain myself, and kissed her softly after her own fashion.' Removing to the chilly courtyard, the couple came outside to see the three journalists off. 'As far as I could see, I saw them standing at the gate waving farewell to me, the brisk winds tossing their white hair,' wrote Nellie.

The fleeting visit, lasting no longer than thirty minutes, made a lasting impression. Jules and Honorine Verne diligently followed Nellie's progress around the globe. When she reached the finish line, a congratulatory telegram was waiting from the Vernes. Front page headlines in the 26 January 1890 issue of *The World* announced the famous author's joy at Nellie's victory. 'Verne's Bravo.' 'The French Romancer In Ecstasy Over the Achievement of "The World's" Voyager'. Quotes from Verne filled the accompanying article by Paris correspondent Sherard. 'She has beaten our friend Phileas splendidly,' Verne said. 'I would have been enchanted to do the same journey, even under the same conditions rushing around the globe not seeing much: would have set off at once and perhaps offered to escort Miss Bly.'

To this day, the American globetrotter who so impressed the French author and his wife returns to the Maison Jules Verne several times a year when an actress dressed in Nellie's travelling garb recounts the story of her thirty minutes with the Vernes.

The unexpected detour to Amiens put Nellie's chances of beating Phileas Fogg in great jeopardy. But like most of the other risks she took, it became a triumph.

Chapter 4

In Which Nellie Gets Back on Track

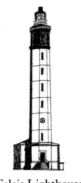

Calais Lighthouse

Calais
22–23 November 1889

'To all kind friends in America, good-bye for the present. It was sharp work to get this train, but I got there all the same.'

Nellie Bly

As the night train whirled from Amiens to Calais, the special moments of Nellie's visit with Jules and Honorine Verne played in her mind. With fondness she recalled their welcome, their encouragement, their conversations, and their farewells as they waved *au revoir* until her carriage was long out of sight. The greeting from their dog Follet, the glow of the fire, the famous author's humble study, the unexpected kisses from Mme Verne: all these scenes unrolled with the rhythm of the steam train as it skirted the coast to Calais. Even without a common language, in a very short space of time, in a country she did not know, a camaraderie had been forged that she already cherished. 'All the way to Calais,' wrote Greaves, 'Nellie did nothing but talk about the motherly kindness of Mrs. Verne.'

The pair were passengers on the exclusive Club Train introduced that year by Compagnie Internationale des Wagons-Lits, operators of the

Orient Express. The concept of a Club Train intrigued Nellie. At first she feared it was the equivalent of a private club and was reluctant to travel on a 'train devoted to men' until she saw some women onboard. Her enquiries revealed that they were travelling on 'the finest equipped train in Europe' with a reputation as the 'pride of France'. Even so, the carriages were so narrow that this elite train seemed more like a toy to Nellie. Once in the dining car, she was impressed by both the service and the food. A full *table d'hôte* menu that finished with cheese and salad was served before guests adjourned to the drawing room carriage for after-dinner coffee. 'I thought this manner of serving coffee a very pleasing one, quite an improvement on our own system and quite worthy of adoption,' Nellie wrote.

When the Club Train pulled into Calais just after 11.30 that night, there were two hours to spare before Nellie's next connection, the weekly India Mail Express train to Brindisi that would sweep her across Europe. She had acquired one of the limited reservations for the Pullman Palace sleeping car on this train that was otherwise entirely devoted to conveying the post. With just a few fellow passengers, she would travel alongside a thousand mail bags packed with letters and newspaper bags bound for India and Australia. The agent for the sleeping-car company in Calais kindly offered to open the Pullman in advance so that Nellie, who had been awake for nearly two full days, could catch up on her sleep. She declined. 'This is the first time since I started when I really felt I had an hour to call my own, or was without anxiety as to whether I could catch this particular train,' she told Greaves. 'I want to look around a bit.'

The night was swathed in black as Nellie and Greaves set off on a walk around the port. Beams cast by the Calais lighthouse painted the sea and sky with ribbons of light. 'Like the laths of an unfinished partition,' Nellie wrote of the patterns playing out around them. The beams brushed so close that she felt she needed to dodge them. Their radiance enthralled her; she wondered if the people of Calais could ever see the stars. As they strolled along the pier at midnight, the two journalists glimpsed a new moon, a glowing eyelash that seemed to wink between the streams of light. It was the same pier that had been captured in oils by British artist J. W. M. Turner in one of his great early works. *Calais Pier*, based on Turner's own experience of arriving at the French port in 1802, depicts the drama as an English ferry boat attempts to dock amid angry, swirling seas. A shaft of sunlight escapes from the menacing clouds above, illuminating the vessel's sail. Passengers grip the boat's hull as women busy cleaning fish on the wave-thrashed pier brace themselves against its railings. But this night, there was no drama, no waves and no people. The Channel was as smooth as black velvet. The pair wandered along the port towards the lighthouse, 'one of the few coast lights worth looking at,' in Greaves' view. To Nellie, the soaring octagonal tower was 'one of the most perfect lighthouses in the world, throwing its light farther away than any other'.

The lighthouse, operating since 1848, still casts its beams out twenty miles to oversee the busiest shipping lane in the world. Resembling a rocket ship ready for blast-off, its white-washed barrel is topped by a black sleeve and decorated with a red star, a symbol for celestial navigation. When Nellie and Greaves were in Calais, the red brick torso of the lighthouse remained unpainted. A lighthouse keeper lived in the substantial house that still stands at its base. Indeed, they had a chat with the resident keeper on duty that night.

Today the automated lighthouse, a heritage-listed French monument, welcomes visitors who pay €4.50 ($5) to climb 271 steps up a spiral staircase for a 360-degree panoramic view. As they circle round the top, visitors can see Calais and its vicinity, the port facilities and the Straits of Dover. On clear days, the White Cliffs of Dover that define England's southeast coast are visible. I have never visited the lighthouse or even seen its beams at night. Like Nellie and most travellers, I only come to Calais to travel somewhere else on the continent.

The port area at Calais, where Nellie and Greaves strolled freely, is now a fortress. White wire fences stretching more than sixteen feet high, topped with coiled razor wire, run alongside the train tracks and motorways approaching the port. CCTV cameras and armed police, often with guard dogs, patrol this 'Ring of Steel' where asylum seekers and migrants have gathered since 1999 in hopes of crossing the Channel to seek sanctuary in the United Kingdom. The French and British governments, along with the European Union, have united to prevent them. Hundreds, at one time thousands, of vulnerable people wait here. Row upon row, mile after mile, the fences are a visible reminder of our inhumanity. I am betting that Nellie, if she were here now, would be tackling the refugee crisis today just as she did the injustices in her day, with her pen and unstoppable resolve to right the wrongs. I use my pen in London to help asylum seekers, refugees and migrants learn the English language. You cannot have a voice if you do not know the language. It is the fastest route to integration.

These days at the Port of Calais, armed French border police search our vehicles for fleeing refugees before we can proceed to the docks. Once through, we are instructed to remain in or near our cars. From the boarding queues, arranged like abacuses around the outbound ferries, I see the lighthouse rising over the industrialised docks. Once onboard, as the Dover-bound ferry manoeuvres into the Channel, I head onto the deck to watch it disappear.

*   *   *

After exploring the bleak French Channel port with Greaves, Nellie admitted that 'there are pleasanter places in the world to waste time than in Calais.' As the 1.30 am departure for the train to Brindisi drew near, Nellie and Greaves headed back along the pier to the monumental Gare Maritime Calais, opened amid great fanfare just five months earlier by French President Sadi Carnot. 'A very fine railway station, generous of size' wrote Nellie of the impressive *gare maritime*, now reduced to a modest ferry waiting area, a fraction of its former enormity. Inside the two journalists had a quick bite to eat before Nellie boarded her fourth train in two days. The journey to Brindisi would take forty-eight hours. There she would swap steam train for steamship and board the SS *Victoria* for the third major leg of her race around the world.

Nellie bade farewell to the escort who had so capably accompanied her through the last three frenzied days. Tracy Greaves waved Nellie off before starting his return journey. As her train pulled out of the station, Nellie slipped a message to him through the carriage window. It read: 'To all kind friends in America, good-bye for the present. It was sharp work to get this train, but I got there all the same.'

*   *   *

I could not follow Nellie's trail from Italy through the Middle East. Stern advice from the Foreign Office and a pledge to my family not to take unnecessary risks meant that potential hot spots of terrorism and war, like the Sinai Peninsula and Yemen, were off-limits to me. I will catch up with Nellie in Ceylon.

Chapter 5

In Which Nellie is Delayed in Colombo

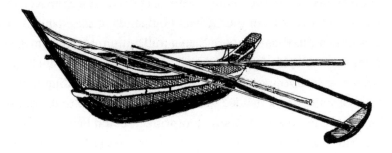

Ceylonese outrigger, Colombo

Colombo, Ceylon (Sri Lanka)
8–14 December 1889

'With all our impatience we could not fail to be impressed with the beauties of Colombo and the view from the deck of our incoming steamer. As we moved among the beautiful ships lying at anchor, we could see the green island dotted with low-arcaded buildings, which looked, in the glare of the sun, like marble palaces.'

Nellie Bly

Was it a mirage? After fourteen scorching days at sea, the emerald island shimmering ahead of them looked like paradise. From the decks of the SS *Victoria*, the sun-worn passengers gazed out on a Garden of Eden afloat on a glistening ocean. They could almost feel the coolness awaiting them under the palms and banyans shading the henna-coloured earth ahead. 'The island, with its abundance of green trees, was very restful and pleasing to our eyes,' Nellie wrote. She and the other passengers could not wait to abandon the *Victoria*, to step on *terra firma*, out of the heat, and divest themselves of the ship's motion that would still linger awhile after they disembarked. 'With all our impatience we could not fail to be impressed with the beauties of Colombo and the view from the deck of our incoming steamer,' wrote Nellie. 'As we moved among the beautiful ships laying at anchor, we could see the green island dotted with low-arcaded buildings, which looked, in the glare of the sun, like marble palaces.'

In the distance, above the rolling treetops, they could see the sacred site of Adam's Peak rising from the Central Highlands of the island. Adam's Peak has drawn pilgrims for more than a millennium to pay homage to a holy footprint embedded on its summit. As the SS *Victoria* anchored in the harbour that morning, pilgrims would have been winding their way down after a sunrise trek that led them more than a mile into the sky. Four religions lay claim to the yard-long footprint. It could belong to Buddha or Lord Shiva; or as Christians and Muslims maintain, it could be the first step Adam made after his banishment from the Garden of Eden. Sacred or not, Nellie admired the scenic backdrop this pyramid-shaped mountain provided for Colombo, then capital city of Ceylon. Marco Polo got to the top of Adam's Peak in 1298. Neither Nellie nor I attempt it.

As her shipmates dutifully boarded the steam launch for the short journey to the docks, Nellie spotted a flock of outrigger canoes drawn like magnets to the SS *Victoria's* arrival. Crafted from the forest and lashed with bamboo and coconut twine, these watercraft were the first true sea-going vessels. In a typical Bly-style leap of faith, Nellie hopped aboard the closest outrigger and sped across the harbour. She was the first of the passengers to step ashore on this small teardrop island. The SS *Victoria* had delivered her from Brindisi, Italy at record speed. Nellie was two days ahead of schedule after navigating the Mediterranean Sea, the newly opened Suez Canal, the Red Sea, the Gulf of Aden and the Arabian Sea. At last she had reached the Indian Ocean and Ceylon; she could not wait to reach the shore.

*   *   *

I am seven miles high, looking down on an ocean of luminous peacock feathers. A peach-blush sun signals daybreak as Air Sri Lanka's Flight UL504 begins its descent to Colombo on the first flight of my round-the-world itinerary tracing Nellie's footsteps. I am catching up with her in Asia. Clouds part, revealing jade forests ringed by the golden coastlines of this tropical island just twenty miles south of India. It took Nellie sixteen days to travel from London to Colombo. I arrive in eleven hours. London is 5,415 miles away and so is my 'usual' life. The adventure has begun. Bundled into window seat 30A, midway along the airliner's cabin, I tuck my tablet into my bag ready for our final approach. Like Nellie, I am anxious to get off and get going. A niggling sense of apprehension would prevail if I let it. But Nellie is in charge now; my sense of purpose outweighs my love of spontaneity.

The Airbus pulls into the gate, the cabin door opens. I dash off, race through immigration and bypass baggage claim to be the first in the arrivals hall at Bandaranaike International Airport. It is Sunday 7 September. I am in Nellie's Ceylon and my Sri Lanka. Travelling as light as her means I can hit the ground running, no lost luggage, no lost time, ready for action. But I am forced to put the brakes on. Japanese prime minister Shinzo Abe arrived here just before me. Outside the relative sanctity of the arrivals hall, crowds have assembled to greet the Japanese leader. He is here to announce funding from his government for a new state-of-the-art terminal which will accommodate nine million extra passengers a year. I have flown straight into an official two-day state visit. The soon-to-be expanded international airport is packed. We are celebrating Japan's generosity. This I learn from Kanchana Thanuja of Camlo Lanka Tours, who is collecting me from the airport. I had indulgently, and now thankfully, pre-booked a private pick-up for my journey into Colombo's city centre. With a shaved head and an athletic build, this young but accomplished tourism professional could be mistaken for a wrestler. Kanchana, dressed in black jeans, a blue-checked shirt and a Manchester United cap, tells me we are witnessing the first visit of a Japanese prime minister to Sri Lanka in twenty-four years.

Winding back the time, twenty-four years ago would have been 1990. My last trip to Sri Lanka was a year later, when I was travelling as a journalist for the Save the Children Fund. A different kind of chaos reigned at the airport when I flew in on 2 March 1991. The country's vicious civil war between the government and the Tamil Tigers was escalating. It was morning rush hour in Colombo; my flight had just landed. Sri Lanka's then foreign minister Ranjan Wijeratne was on his way to work in the city centre when a car bomb exploded, killing him and eighteen others. Instructed to get to my hotel and stay put, it was a full day before I could leave it again. I remember my frustration at being confined to a drab, cramped business hotel when a bustling South Asian city awaited me just outside the door. Save the Children had sent me to Sri Lanka to document its night schools for street children forced to work during the day. It was an assignment *à la* Nellie Bly, the ones I like the best. I wanted to get started. After the bombing, I was asked to avoid the war zones in the far north of the country.

Twenty-three years later, I can travel wherever I want in Sri Lanka. My only restrictions are self-imposed, just the places where Nellie went. In 2009, after almost three decades of fighting, a fragile peace settled on the island. Sri Lanka is now in recovery mode. Here at Bandaranaike International Airport, the Japanese prime minister's visit is testimony to the country's determination to help rebuild its economy through tourism. Snaking through the crowds to the parking area, Kanchana leads me to the Camlo Lanka van for what should be the short drive into Colombo. It is further away than I thought; not in miles, but in time. All major thoroughfares are blocked for the Japanese premier's official motorcade and the cultural festivities accompanying it.

Colombo is on parade. A motorcade of black state sedans creeps along the centre lane of the E12 Expressway which has now become a VIP 'red carpet' into the city. Kanchana and I follow in his spotless white Toyota van, windows closed, air conditioning turned up high. The roadsides have become outdoor stages showcasing the talents of every dance studio and music school in the region. I watch their performances through the windscreen. The tropical colours of the national flag unfurl beside us as dancing children wave silky banners dyed saffron, maroon, emerald and orange. Majorettes, draped and belted in lime, lemon and tangerine, twirl snowy plumes in place of batons as barefoot dancers pass bouquets of pale pink lotus flowers to passing officials. As we inch along behind the procession, I open the van window to take in the fruity scent of the lotus blossoms. Hollow beats from hide-covered drums slip inside, along with murmurs from the crowds gathered in their Sunday saris and sarongs to welcome the Japanese delegation. The standard half-hour commute into Colombo expands well beyond an hour. Kanchana is relaxed, so I am too. His genial manner inspires confidence … and patience. Besides, not everyone gets front row seats to a VIP cultural performance as soon as they arrive in Sri Lanka, unless of course they are the Japanese prime minister and his entourage. At last, the ivory colonnades of the five-storey Grand Oriental Hotel come into view. I am ready to be here. It is the only hotel still standing that hosted Nellie on her world journey.

For Nellie, 'a nearer view of the hotel did not tend to lessen its attractiveness – in fact it increased it.' After her dash across the harbour on the outrigger, Nellie had already engaged her accommodations when the steam launch arrived with the other passengers. My own accommodations at the Grand Oriental were arranged a month in advance on the hotel website promoting it as a 'legend yet living', and inviting me to 'feel the colonial times'. I have booked in for the best of both worlds, modern and colonial. My penchant for colonial architecture just about allows me to set aside the implications of power, wealth and status implicit in empire. I am drawn to old-world grandeur. I love how oriental touches magically transform classical European architecture, taking account of the climate with balconies, open-air arcades, latticed windows and shutters that welcome breezes, but moderate the sun. I will see a lot of this architecture on my trip as I follow Nellie to colonial outposts across Asia. But alas, much of the Grand Oriental's colonial splendour is now just a memory. I had so hoped to stay in the same room as Nellie did; but, predictably, records of her late nineteenth-century sojourn no longer exist. I cannot blame the management for not sharing my enthusiasm to find 'all things Nellie Bly' in the one hotel where our paths might cross. Many other Empire-era hotels were completely lost before their architectural significance was recognised.

'It was a fine, large hotel,' wrote Nellie, 'with tiled arcades, corridors airy and comfortable, furnished with easy chairs and marble-topped tables which stood close enough to the broad arm-rests for one to sip the cooling lime squashes or the exquisite native tea or eat of the delicious fruit while resting in an attitude of ease and laziness.' No cooling lime squash for me; but I am greeted with a multi-blend concentrated fruit juice as I check-in. The ice cubes clink as I drink.

Like Nellie, I have five days to spend in Sri Lanka. But it was five days too long for her. She was ready to depart as soon as she arrived. Her next ship, the SS *Oriental*, was waiting to set sail, but it was stuck in port until the mail and passengers arrived from the *Nepaul*, a P&O steamer on its way back from Calcutta. During her exasperating five-day wait for the *Nepaul*, Nellie became well acquainted with the Grand Oriental. In her day, nearly-naked snake charmers and magicians 'colonised' the corridors, along with 'high-turbaned merchants' who snapped open little velvet boxes to expose the 'most bewildering gems'. Deeply-dark emeralds, fire-lit diamonds, exquisite pearls, the lucky cat's eye, 'rubies like pure drops of blood,' Nellie listed them. One of the world's oldest sources of sapphires and rubies, the gemstone trade here dates back 2,500 years. These precious stones, particularly sapphires, spill from the mountains of Sri Lanka, once called the Island of Jewels. As Nellie noted: 'No woman who lands at Colombo ever leaves until she adds several rings to her jewel box, and these rings are so well known that the moment a traveler sees one, no difference in what part of the globe, he says to the wearer, inquiringly: "Been to Colombo, eh?"'

No woman, except of course for Nellie Bly, who seems to have resisted the allure of the gems. My own engagement ring is set with a Sri Lankan sapphire. Nellie's prized ring was a gem-less gold band she had worn on her left thumb since the day in 1888, penniless and almost homeless, she landed her job at *The New York World*. The ring was her talisman and had only been off her thumb 'for three unlucky days'. I tried to replicate Nellie's ring so I too would have a good luck charm for my travels, but my bulbous thumbs refused. 'Why not a piercing instead?' my daughter Acadia suggests on a phone call from her University of Winchester dorm. Thumb ring versus body puncture, I am not sure of the logic. Nose, bellybutton, lips, eyebrows and other unmentionable parts are out; but perhaps a cheeky upper earlobe stud could bring good luck. Acadia says she knows just the place. On my next visit to Winchester, she takes me to Asgard Piercing and Tattoo Studios where we are greeted by a multi-pierced attendant who specialises in tattoos, but that is not an option.

Breaking the skin with a mini paper punch, the piercer inserts a micro titanium rainbow barbell in my upper right lobe. It joins the silver Celtic hoops inhabiting the lower lobe piercings performed by my mother in 1966 using a darning needle and ice. It is not a thumb ring like Nellie's, but I like finding tiny silvery earrings to fill my lucky gap.

* * *

Colombo was the first place Nellie saw American gold coins since leaving home. But she did not expect to see them dangling from the pocket watch chains of wealthy Ceylonese merchants who used them as jewellery, not as currency. In fact, American gold could only be exchanged at sixty per cent of its value, but the diamond merchants of Colombo would pay a high premium for twenty dollar gold pieces. 'The richer the merchant, the more American gold dangles from his chain,' Nellie wrote. 'I saw some men with as many as twenty pieces on one chain.'

According to Nellie's account, most of Colombo's jewellery sales were conducted inside the corridors of the Grand Oriental Hotel. Today the commerce continues outside with no less than seven jewellery shops a stone's throw away from the hotel's York Street premises. Along with the former commerce inside, much of the Grand Oriental's original site has been lost. I am quite sure Nellie's quarters vanished when the original 154 luxury and semi-luxury rooms shrunk to eighty in the 1960s. So did its landscaped tropical garden, the only hotel garden in Colombo. In Nellie's day, rainbow lights illuminated flame trees, ferns, orchids and palms in the garden where the resident orchestra played on Wednesday and Sunday evenings. The orchestra also performed during *tiffin* (luncheon) and dinner in the hotel's dining room. You could say the tradition carries on at the Grand Oriental; the resident orchestra is no more, but pianists and saxophonists still serenade guests during lunch and dinner buffets.

Nellie raved about the fine food, service and decor in the Grand Oriental dining room which matched 'the other parts of the hotel with its picturesque stateliness'. Here she sampled her first curry. Spooning shrimp over boiled rice and topping it with chutney, chillies and specks of Bombay duck, Nellie loved the fiery aromas of cardamom and ginger rising from her plate. She declared her first curry 'very unsightly, but very palatable'. The South Asian dish became her first choice at meals until she found that after a particularly hearty meal, 'curry threatened to give me palpitation of the heart'. Nellie and the other diners were fanned during meals by embroidered *punkas*, long strips of cloth fastened to bamboo poles suspended from ceilings. The *punkas* were linked by ropes, and drawn back and forth by servants before the advent of electricity and ceiling fans. Derived from the word *pankh*, the wings of a bird which produce a draft when flapped, *punkas* were common across South Asia, and on ships travelling in the East, according to Nellie. 'They send a lazy, cooling air through the building, contributing much to the ease and comfort of the guest,' she wrote.

My comfort is enhanced by the overly enthusiastic central air conditioning that long ago replaced the punkas at the Grand Oriental Hotel. It is another era, and even though the hotel's glory is not what it once was, the staff here is fiercely proud of its legacy. On the walls of the saffron and cinnamon-toned lobby, vintage photographs and advertising flyers attest to the Grand Oriental's heritage as 'the oldest and best known first class hotel in Colombo,' stating that 'its excellence ... is too well-known to need description'. One framed advertisement boasts of the 'Best Modern System of Drainage, Hot and Cold Water to Bed and Bathrooms'. Potential guests were invited to apply to the manager for terms.

A five-night stay here by Russian writer Anton Chekhov from 12–18 November 1890 lives on not only in a bronze bust fixed to a marble plinth in the lobby; but also upstairs in a hotel suite named after him: the Anton Chekhov Memorial Room, suite 304. Chekhov is said to have been writing his short story *Gusev*, about the death of a soldier on a sea voyage, when he was here. To him, Ceylon was paradise on Earth.

There is absolutely no mention of Nellie Bly's five-day visit in the lobby, or anywhere in this hotel that hosted her longer than any other on her iconic voyage. Her visit preceded Chekhov's by eleven months. In an attempt to help me place Nellie here, the front desk staff tell me of a tiny museum on the fourth floor. Maybe there will be a trace of her inside. But it is not until my last night at the Grand Oriental that I finally gain access to this perpetually locked and virtually secret museum. And that is only because the front desk staff pleaded my case with their manager. Heading up at last, I am led by Chandika, assistant front officer manager, Chaminda, bell hop and Dushaatha, security officer, who all seem as anxious as I am to get inside this one-room museum. Bundling all four of us into the vintage elevator, said to be the oldest in Sri Lanka, is good preparation for packing ourselves into the mini 'museum'.

We can admire, but not touch, the historic items, Chandika tells us as he unlocks the door. A musty cloud spills into the hallway, escaping from the airless room that is not much larger than a loo. Edged into the back wall, a brown *étagère* displays tarnished silver plates and tea pots, a classroom wall clock, a white porcelain soup tureen, a copper coffee pot, and three sets of salt and pepper shakers emblazoned with the hotel's former logo, a gargantuan elephant flanked by palms and inscribed with 'G.O.H., premier hotel'. Atop the *étagère*, twenty-six short-stemmed pewter goblets line up two-by-two below brass statuettes of Sinhalese deities. The rungs of louvred shutters serve as picture rails for faded photographs in fluted wooden frames. The images are obscured by the light of a single fluorescent light; we have to squint to see them. Most portray this colonial dowager in her days of glory; but there are also vintage snaps of local tourist attractions, portraits of former staff, and thirty-three other photographs that have nothing to do with Nellie Bly. Clearly it is up to me to bring her back. Move over Chekhov.

*   *   *

I had heard that Sri Lanka was renowned for its hospitality. I experience it big time. Not only in Sri Lanka, but in London where a Sri Lankan friend of a friend introduced me by email to his closest pals in Colombo. That is all it took for Lakmini Raymond and her two millennial sons Jévon and Devin to share their time, their local knowledge and their friendship with me. I was not surprised to learn that Lakmini is a leader in the Colombo hospitality industry, serving the best hotels in the area. She clearly takes her work home with her, making this stranger feel very welcome just hours after I arrive in her country. Lakmini is a fashionista, frequently appearing on the list of Colombo's best-dressed. Today she is attired entirely in white – linen trousers and a muslin blouse delicately embroidered with blossoms and stars. I am clothed in what I will wear for most of my trip: khaki travel skirt, t-shirt and bandana.

Lakmini meets me at the Grand Oriental. She has a plan that starts ceremoniously with a cup of legendary Ceylon tea at the former Old Colombo Dutch Hospital nearby. Built in the late seventeenth century, after the Portuguese left and before the British arrived, the red-roofed former hospital wings are now home to a dynamic mix of shops, bars and eateries. It is the Sri Lankan equivalent of London's Covent Garden, Boston's Faneuil Hall and San Francisco's Ghirardelli Square. After a night on a plane and a morning in a car, the flowery tang of this large-leafed, high-grown Ceylonese tea is a rejuvenator. I am experiencing two Sri Lankan icons at once: the famous tea and an outstanding heritage site. Lakmini, who never even knew me an hour ago, has arranged a dinner for the newly inaugurated Nellie Bly trail team that also includes Jévon, Devin, and her friend Jagdesh Mirchandani, who she calls 'Sri Lanka's Wikipedia'.

In the Curry Leaf Restaurant at the sky-rise Hilton Colombo, we are seated before a full-length plate glass window looking onto an island of floodlit fountains rising from a satiny black lagoon. We can hear the music of the tumbling water. Sprinkled with fairy lights, the outside is almost inside. Sweet, peppery aromas drift our way from the buffet nearby. We order curries from the menu, shrimp for me with Nellie's first curry in mind. In the distance, a group of dark-suited men marches by, cords spiralling from their ears like ringlets. 'Security guards,' Lakmini explains, 'the Japanese prime minister is staying here.' First the airport, now the hotel. It feels more like I am following Shinzo Abe than Nellie Bly. Jagdesh, a professorial type with soft silver hair, and a white beard and moustache, pulls out his spectacles to study Nellie's own account of her journey. I always carry a print-out of her book *Around the World in Seventy-Two Days* with me. Chapter IX, 'Delayed Five Days' and Chapter X, 'In the Pirate Seas' cover her time in Ceylon. I am not sure which of Colombo's thirty-plus temples Nellie visited, but the team recommends Gangaramaya, complete with a sacred Bodhi tree, as one of the most important. With a list of Nellie Bly must-sees, together we construct my itinerary for the next day.

I awake to Binara Full Moon day across Sri Lanka. It is a poya, a Buddhist public holiday, when businesses close, alcohol and meat are forbidden, and women in white flock to temples. The Sri Lanka *Daily News*, which I am reading over a bowl of fruit in the hotel's Harbour Room restaurant, has a full report. It is 9.00 am and I am gazing at the port where Nellie arrived at precisely the same time on 8 December 1889. Today's Binara Poya commemorates the value of women in the Buddhist religion, says the *Daily News*. This poya marks the founding of the Bhikkhuni order of ordained nuns established by Buddha in the Third Century, the first time that women were accepted in monastic life.

The order died out five centuries later. After a thousand years, the order has recently been revived. Today there are more than 1,000 Bhikkhuni nuns across Sri Lanka, according to the newspaper. I later discover that the revival reaches as far as England, with the opening of the United Kingdom's first residence for Bhikkhuni nuns in Oxford in 2018.

Here in Colombo, on this holy day for females, the Buddhist Temple of Gangaramaya is drawing far more devotees than tourists to its sacred site beside the green waters of Lake Beira. Women of all ages, sizes and backgrounds, adorned in white saris and skirts, gather to celebrate their day. Like a long white serpent, they wind around the out-stretched branches and vine-draped roots of the holy Bodhi tree. Lord Buddha attained enlightenment under such a tree. I am obtaining photographs. In an otherwise white throng, I stand out in a fluorescent lime t-shirt and fuchsia bandana. At least they reflect the tropical colours of the lotus blossoms offered by devotees to the colossal yellow Buddha seated in the main sanctuary. Outside, the earthy scent of sandalwood wafts from silt-filled pots embedded with incense. Gangaramaya is the first of many temples I will visit on this journey; the Temple of the Tooth in the sacred city of Kandy is the second. Nellie found temples of 'little interest' throughout her trip, but they captivate me. However, we both adore fine hotels.

The 'smoothest, most perfectly made roads' Nellie ever saw led to the exclusive Mount Lavinia Hotel, eight miles outside of Colombo. They were red, she noted, and constructed by convicts. 'Many of these roads were picturesque bowers, the over-reaching branches of the trees that lined the waysides forming an arch of foliage above our heads, giving us charming telescopic views of people and conveyances along the road,' she wrote. She called them tree-roofed roads. Crafted from jungle palms, the thatched huts framing the roads in Nellie's day have given way to discount shopping malls, burger chains and car dealerships. They choke both sides of the road, western-style. But all traces of this global commercial congestion vanish as the early nineteenth-century Mount Lavinia Hotel comes into view. Nellie described the Mount Lavinia as a 'castle-like building glistening in the sunlight … on a green eminence overlooking the sea'. A single guard stands before this irresistible 220-year-old colonial hotel. Except for his shiny black shoes, belt and braided epaulettes, the guard here is clad entirely in white, just like the heritage hotel who employs him. His knee socks climb towards Bermuda shorts topped by a long-sleeved safari-style jacket. Best of all is his tropical pith helmet capped with a silver point. I cannot resist posing for a photograph with him.

I am here at one of Sri Lanka's oldest hotels, the 275-room Mount Lavinia, courtesy of new friends and newlyweds Stephanie and Moahan Balendra. Like Lakmini, we are connected via emails from London. They have allocated their day off, and navigated holiday traffic on the notorious Galle Road, to introduce me to this celebrated hotel. They have now, inadvertently, but with enthusiasm, joined the Nellie Bly research team. The newlyweds decline my offer to take their picture with the obliging guard, but kindly snap mine before we enter the gleaming white lobby with gold lighting. Building on its heritage, the hotel's interiors feature life-size curios from times gone by including an ancient rickshaw, a two-wheeled, human-powered Asian mode of transport that Nellie encountered for the first time in Ceylon. It troubled her initially, but not for long. 'I had a shamed feeling about going around town drawn by a man,' Nellie confessed. 'After I had gone a short way, I decided it was a great improvement on modern means of travel; it was so comforting to have a horse that was able to take care of itself,' she decided. Stephanie, taking hold of the antique cart's long wooden handles, leans forward and assumes the role of a rickshaw driver so I can take her picture. Rickshaw drivers in Nellie's day wore little more than a sash and a large mushroom-shaped hat. Stephanie, a real estate and finance expert, is dressed in jeans and a sleeveless navy top. Her husband Moahan, an attorney-at-law, sports tangerine shorts and a plaid cotton shirt.

She and Moahan were not married here, but Stephanie explains that the Mount Lavinia is a favoured spot for weddings. This is partly due to its exotic location crowning a promontory over the ocean, partly for its exquisite colonial elegance, but mostly for its deeply romantic legacy. The Mount Lavinia is the setting for the love story of Ceylon's second Governor Sir Thomas Maitland and the exotic dancer Lavinia Aponsuwa. Unhappy with the accommodation provided to him by the British government, Sir Thomas constructed the current hotel in 1806 as the Governor's Residence. It included a tunnel that stretched from a deserted well near Lavinia's home through to the cellars of the Governor's mansion where the two lovers could meet. Their romance continued for six years until Sir Thomas was called back to Britain. He is said to have named the residence for Lavinia. The tunnel remains to this day, along with the legacy that the Mount Lavinia calls 'the legend that everyone wants to believe'.

Above the tunnel, just steps from the hotel, landscaped gardens give way to the beach. We wander over for a seafood lunch on this coastline known as Paradise. Wooden tables lodged in cappuccino sands are sheltered from the sun by a canopy covered with plaited coconut palm fronds. Hurricane lamps, attached with twine, sway from the beams. We choose a table closest to the water where a feisty Indian Ocean slaps the shore. A single outrigger is wedged in the sand nearby; pale blue paint peels from its banana-shaped hull. A fisherman in an indigo-striped sarong is adjusting the ropes inside. It is just like the outrigger that Nellie rode to shore when she arrived in Colombo.

Further down the coast in Colombo, the Galle Face Hotel, known by Sri Lankans as the 'grand old lady', gracefully anchors one end of Galle Face Green, a swathe of ocean-side lawn attracting joggers, kite-flyers, picnickers and sunset gazers. Built in 1864, it vies in old-world charm with the Mount Lavinia. In Nellie's day, the Galle Face was known as the best hotel east of the Suez Canal. She was there on a 'sweet, dusky night' soon after she arrived in Ceylon. Very rarely did Nellie let go of her fixation with the race long enough to relax, especially when she was delayed. But the enchanting coastal setting of the Galle Face Hotel seemed to take the edge off her frustration over the long-awaited arrival of the SS *Nepaul*. Reclining on a chaise lounge under the eaves of the hotel's veranda, Nellie could see 'where the ocean kisses the sandy beach' through a forest of tall palms that has since diminished. She listened to the 'music of the wave', its 'deep, mellow roar'. And she allowed herself to 'drift out on dreams that bring what life has failed to give; soothing pictures of the imagination that blot out for a moment the stern disappointment of reality.'

I am faced with the 'stern disappointment of reality' at the Galle Face Hotel when I arrive around twilight. One of the finest Colonial hotels in the world looks more like a building site. This 'grand old lady' is in the midst of a facelift. Massive hoardings engulf her elegant façade, shielding the extensive restoration work underway. Nellie's veranda is also off-limits, so I stroll to a table on the beach. A waiter dressed in a crisp white uniform with burgundy epaulettes immediately arrives to take my order and place a tealight where I sit. Nellie loved the cooling lime squashes served at the Galle Face and the Grand Oriental, but I order a glass of Sauvignon Blanc. Just as the hoardings hide the hotel from view, dove grey clouds blanket the horizon obscuring what I can only imagine as a glorious sunset. Even so, Nellie's mellow 'music of the wave' is playing. A sun-bleached outrigger is silhouetted against the sea as the evening breeze carries the scent of salt across the sands. I can taste it on my tongue. Tomorrow I will follow Nellie to the sacred city of Kandy, far from the coast, up into the island's Central Highlands.

Chapter 6

In Which Nellie Travels to Kandy

Temple of the Tooth, Kandy

Kandy, Ceylon
10 December 1889

'Kandy is pretty, but far from what it is claimed to be.'
Nellie Bly

Nellie took the 7.00 am train from Colombo to Kandy and so do I. She opted for a day trip; I will spend the night. Colombo Fort station is heaving. Train departure announcements echo across the station in Sinhala, Tamil and English, but the words get lost in the rush hour pandemonium. Abandoning all courtesies, I roll through the multitudes like a tank; head down, no eye contact. I am in Nellie Bly race mode until I finally reach platform two, where the train to Kandy awaits, humming on steel rails. As soon as I arrive, an early morning monsoon bursts full force through the Colombo skies. Raindrops ricochet off the platform's faded pink pantiles as the downpour sweeps the station. The rain is warm and smells of mushrooms, but I remain dry under the metal-roofed canopy until it is time to board the train.

Inside the imperial red Intercity Express to Kandy, seat numbers are neatly stencilled in white paint above the windows. I feel lucky to have snapped up a first class seat in the air-conditioned observation car; it is well worth the extra 200 rupees (£2/$2.50) to travel in a roomy carriage with this much glass. The railway carriages harken back to Nellie's era, except for the reclining seats upholstered in screaming blue velour showered in comets and stars. I am on my way to the heart of the island, the Central Highlands, crowned by the sacred city of Kandy.

As we depart Colombo, shanties forged from scrap metal and discarded tyres huddle near the tracks. Children wave as we pass. Gritty urban scenes gradually surrender to nature as banana trees, coconut palms and blossoming vines emerge from the auburn earth. Further along the tracks, tall grasses, tangled vegetation and narrow-trunked trees meld into an impenetrable landscape, a jungle inhabited by chestnut monkeys and black eagles. As we snake along broad-gauge tracks carved through the rock two centuries ago, we are at one with this iron horse; it pants with exertion as the wheels pummel the rails. We feel its every move, swaying and bouncing in our seats as if they were saddles. It is a train dance to Kandy, the island's second largest city and a UNESCO World Heritage Site. Distant hills begin to shape the horizon.

Minor confusion over the stencilled seat numbers brings the Stapels family from Germany into my life. We sort out our seats, and before the journey has ended, I am invited by Bernie to join his wife Redda and their teenage daughter Julia for a private sightseeing tour of Kandy. At first I hesitate, keeping in mind I have a specific itinerary for tracking Nellie and only thirty-six hours to do it. But I push her aside and accept with pleasure. In the end, Nellie steps in anyway and the day includes a visit to the Peradeniya Royal Botanic Gardens where she spent several happy hours. What luck to be personally delivered to places I want to see, and to experience it all with the Stapels. They have two weeks to explore the island; Kandy falls midway in their trip. Bernie, tufted grey hair and mini-sideburns, has organised this holiday for his two girls with Teutonic precision. His daughter Julia carries a 35mm camera, rather than a mobile phone, to document her family's travels. Her tawny hair is braided into a single plait that just about reaches her waist. Redda's hair, captured in a ponytail, is just as long and lovely. In true tourist mode, Redda is covered in elephants marching in rows along her turquoise sarong, up to her v-neck t-shirt featuring a mother and baby linking trunks. The German-speaking guide they have engaged awaits us on the platform and ushers us into a minibus. When he can slip them in, the guide shares a few facts in English for me.

Just outside of Kandy, rows of glistening bushes blanket the hillsides at the Geragama tea plantation, one of the island's oldest. Nellie did not visit a tea plantation when she was here, but I do courtesy of the Stapels. It is our first stop. Ten national flags are planted along the tea factory's four-storey corrugated iron façade; the Union Jack and German tricolour are among them. We are here to see how one of the finest teas in the world is processed from hand-plucked shiny leaves into loose tea and tea bags. We follow our guide and watch how tea leaves are dried, fermented, rolled, graded and sorted. The Stapels family stocks up on tea in the factory shop; but travelling light like Nellie means there is no room in my small bag for even one box of authentic Ceylon tea.

The roots of Ceylon's tea industry were cultivated at our next stop, the Peradeniya Royal Botanical Gardens, where I rejoin the Nellie Bly trail. The island's first tea bushes were planted and tested in these grounds sheltered by mountains and bounded on three sides by the River Mahaweli, Sri Lanka's longest. These revered lands date back to 1371 when Kandy was a kingdom and the site was a pleasure garden. Some 450 years later, in 1821 under British rule, Peradeniya became a botanical garden. Not in the least impressed by other sites in Kandy, indeed 'disgusted with all we found worth seeing', Nellie admired the 'great botanical garden' which she said 'repaid us well for our visit'. Us too. After being confined on a train and in a tea factory, we dive out of the minibus to stroll among the orchids, spice gardens and greenhouses recommended by our guide. We follow a continuous border of ruby, amber and jade bedding plants along the lanes that traverse the gardens. Julia locates the black bamboo grove where we pose with soaring ebony stalks wider than our hands. Doe-eyed couples wound round each other like ribbons occupy benches in scenic spots. A double row of towering Royal Palms casts sun stripes on the avenue they border. Their elephant-grey trunks shoot sky-high, exploding into cascades of fronds inhabited by fruit bats. A fleeting rain shower unleashes the zesty, sweet scents of cinnamon, cardamom and vanilla as we shelter in a stone-built gazebo with a sombrero straw roof. These gardens boast 300 species of orchids. Most are inside the Orchid House where they hang like necklaces, row upon row, in exquisite colours and flamboyant shapes: starbursts, slippers, ballerinas, ghosts. Between the endless lines of exotic specimens, a gardener balances a hideous black scorpion the size of a glove on his arm. Poised for action, its spiny barbed tail curls towards castanet pinchers ready to grab. We do not hang around.

As with most excursions, our guided tour ends in an over air-conditioned emporium offering local handmade crafts and souvenirs: carpets, carvings, wall hangings, clothing. Redda and Julia are being schooled in the art of sari swirling by an enthusiastic saleswoman who provides them each with a Hindu *bindi*, a black dot to place between their eyebrows. They pose together in pink-embroidered saris, placing their hands in the customary *namaste* greeting, palms touching and fingers pointing upwards. I photograph them before saying thank you to each family member, especially Bernie as my host, and take my leave.

The Queen's Hotel in Kandy, like the Mount Lavinia near Colombo, is the former home of a British Governor in Ceylon. Tonight it is home to me. Steps away from Kandy Lake, this four-storey heritage hotel stands as a visible 'almanac' of British colonial architecture. Defined by balconies and balustrades, the hotel is topped with dormers that could have been cut from pinking shears. A pointed cupola, resembling an oversized pith helmet, rests atop a bell-shaped dome that crowns the principal entrance. Queen's could be where Nellie took her luncheon on the day she was here. It is just her kind of place.

The nearby Temple of the Sacred Tooth was not. Nellie dismissed one of the Buddhist world's most sacred shrines in less than thirty words: 'In one old temple, surrounded by a moat, we saw several altars of little consequence, and a piece of ivory that they told us was the tooth of Buddha.' Tell that to the multitudes of devotees and tourists that flock to the temple every day to venerate the upper left canine tooth of the Lord Buddha. I join the legions for the 6.30 pm service, the Evening Thevava, when the Sacred Tooth Relic shrine door will be opened. As I stroll from the Queen's Hotel, the peaked roofs and pinnacles of the vast white temple complex come into view. Waist-high rows of scalloped stone walls, honeycombed with small shapes, wrap the temple buildings in a lace trim. An elevated octagonal pavilion capped with a terracotta 'hat' and circled by a copper-coloured moat, oversees the one-time royal palace grounds of the former Kingdom of Kandy, capital of Sinhalese kings from 1592 until 1815. The last king, Sri Wickrama Rajasingha, addressed his subjects from here.

My $10 (£7.50) entrance fee will allow me to tour the temple and join the evening service to venerate the tooth relic. But not quite yet. First I must pass through a lengthy inspection line to ensure I am properly attired. I am barefoot; my arms are covered and so are my legs, at least as far as my knees. I checked the rules before I came. But it is not sufficient. Without warning, an attendant puts her hand in my bag, whips out my shawl, wraps it around my middle, and transforms it into a sarong. Now I will not offend. Slightly bewildered, I join white-robed pilgrims clutching bouquets of lotus blossoms and frangipani, and baskets of tropical fruit purchased at the entrance, to present as gifts. Crashing drumbeats shatter the crowd's murmur, signalling the approach of the ceremonial hour and the time to move into the *Pallemaluwa*, the ground floor, to climb the stairs to the upper hall which eventually leads to the sacred inner chamber. As we enter through vaulted corridors under heavenly painted ceilings, the relentless hammering of drums shifts into a raw, rhythmic beat. We are now before the Drummers Courtyard where a band of men wrapped in white sarongs and scarlet cummerbunds strike double-sided drums made of wood, rope and hides. One of them is playing an oboe-like instrument that sends shrill, nasal riffs high above the drums, like a snake charmer enticing a serpent to rise. The crowds and temperature intensify with the deafening music. We are shuffling through a stifling haze of flaming coconut oil, smouldering incense and perspiration. Nellie was right. I wish I was not here; a slight wooziness is setting in. But it is too late to exit this sluggish queue. At last I see pilgrims depositing their fruit and floral offerings on long tables. We must be close. Eight minutes later it is my turn to view, not the tooth relic, but the priceless golden stupa-shaped repository in which it resides – seven gilded, gemstone-encrusted caskets nested together and gleaming from a distance. One brief glance and I am moved on, my veneration of the Sacred Tooth Relic is complete. At least Nellie was able to see the Lord Buddha's tooth, even if she was not convinced of its authenticity.

Outside I clear my head, inhaling the evening air. The temple compound is bathed in a dusky pink as I wander along Kandy Lake back to the Queen's Hotel. Tomorrow, before catching the train back to

Colombo, I will return to the lake to look for the indigenous monkeys, egrets and storks that make their homes here.

'Kandy is pretty, but far from what it is claimed to be,' Nellie wrote of her day trip here. I do not agree with her. Lodged in a forested valley commanded by mountain ranges, the sacred city of Kandy and its rich heritage is a highlight for me. Nellie's disappointment surely sprang from the stress derived from the agonising delay to her journey. By now, even with the two days she had saved upon arrival, she was really pushed for time. She was so keen to get on with the race that nothing could please her, not even a tropical island paradise.

Back in Colombo, the blackboard in the Grand Oriental's corridor at last announced that the SS *Oriental* would sail for China the following morning at 8.00 am. Nellie was up at 5.00 am. 'I was so nervous and anxious to be on my way to China that I could not wait a moment longer than was necessary to reach the boat,' she wrote. In any case, she had to wait. Until the SS *Nepaul* arrived, the *Oriental* could not depart. The first passenger aboard the *Oriental*, Nellie found the ship deserted except for a 'handsome, elderly man accompanied by a young blonde man in a natty white linen suit' who were promenading on the deck. They were the ship's engineer and the ship's doctor. Asking when the *Oriental* would sail, Nellie was informed that the *Nepaul* was still nowhere in sight. 'Waiting for the *Nepaul* has given us this five days' delay. She's a slow old boat,' the engineer replied. Nellie snapped. 'May she go to the bottom of the bay when she does get in,' she said, calling the steamship 'an old tub'. With a twinkle in his eye, the engineer mentioned that Colombo is a pleasant enough place to stay. 'It may be if staying there does not mean more than life to one,' she replied. She knew that nobody could possibly understand what this delay meant to her. Nellie conjured 'a mental picture of a forlorn little self creeping back to New York ten days behind time, with a shamed look on her face and afraid to hear her name spoken'. The thought of it made her laugh out loud. Nellie had a sense of humour; she may have also had a premonition. Less than a year after she kept Nellie waiting, the SS *Nepaul* did end up at the bottom of a bay, but not the Bay of Bengal as Nellie wished. On 12 December 1890, the *Nepaul* became one of the largest vessels ever to sink in the Plymouth Sound, a bay off the English Channel in Devon. There was no loss of life as the steamship went under, but most of the cargo was lost. What little remains of the *Nepaul* still lies on the floor of the Sound. Her anchors, winches, steel plates and frames, along with china and cutlery with the P&O crest, have been spotted by divers.[1]

Shortly after her outburst before the *Oriental's* engineer and doctor, Nellie spotted a pillar of smoke just above the horizon. The *Nepaul* at last. The *Oriental* sailed at 1.00 pm. 'I found it a relief to be again on the sweet blue sea,' she wrote, 'out of sight of land, and free from tussles and worry and the bustle for life which we are daily, even hourly, forced to gaze upon on land. Only on the bounding blue, the grand, great sea, is one rocked into a peaceful rest at noon of day, at dusk of night, feeling that one is drifting, not seeing, or caring about fool mortals striving for life'.

*   *   *

Nellie travelled on board the *Oriental* of the P&O Steamship Line to Penang, Malaysia where she spent a day visiting a waterfall, a Hindu Temple and Chinese shrine. I will catch up with her in Singapore.

Chapter 7

In Which Nellie Makes It Half-Way Round the World

McGinty the monkey

Singapore
18 December 1889

'If I fail, I will never return to New York. I would rather go in dead and successful than alive and behind time.'

Nellie Bly

Nellie steamed into a quintessentially colonial Singapore. Starting as a British trading post in 1819, it had already been part of the British Empire for seventy years when she arrived. Winged junks and sandal-shaped sampans shared the cluttered harbour with steamships. Bumboats, eyes painted on their bows so they could 'see', transported their cargos past stilt houses at the water's edge. Grand neoclassical buildings lined streets brimming with rickshaws, bullock carts and horse-drawn carriages known as gharries. It was the crossroads of the East, where the Indian Ocean meets the South China Sea.

I fly into a self-governing, 'Disneyfied' Singapore sealed in a western veneer. Changi Airport, voted the best in the world, feels more like a Club Med, a mini package holiday for passengers who can take a dip in rooftop swimming pools, watch free movies twenty-four hours a day, park their kids in sky-high playgrounds, and explore butterfly and orchid gardens. A model of efficiency and a microcosm of Singapore itself, Changi is one of Asia's busiest airports. Today's Singapore corrals street food peddlers into hygienic food courts called hawker centres, hosts nightly sound and laser shows, and enforces harsh laws that forbid chewing gum, littering, un-flushed toilets and same-sex relations, all punishable by fines or imprisonment. Capital punishment is alive and well here.

Singapore, the world's third most densely populated country, is a monument to order, consumerism and futuristic architecture. But hidden under its ultra-modern mask, vestiges of the island that Nellie saw endure. Bundled between skyscrapers, graciously restored Victorian government buildings now devoted to art and culture speak of Nellie's era. Away from colonial grandeur, architecture was shaped from indigenous materials and reflected local cultures, styles and the surrounding environment. Singapore was raw and exotic, not shiny and synthetic. It was, and still is, a melting pot of races, cultures and religions occupying the world's only landmass classified as an island, a city and a country.

When she reached Singapore on 18 December, Nellie was half-way round the world. Just eighty-eight miles from the equator, she was at the southernmost point of her journey. She was also at the end of her patience. By the time the SS *Oriental* arrived in the harbour, darkness had descended on this island less than half the size of London. It was too late and too treacherous to risk docking at the quay. They had no choice but to anchor in the harbour until morning. Nellie was livid. 'The sooner we got in, the sooner we could leave,' she wrote. 'Every hour lost meant so much to me … I was wasting precious hours lying outside the gates of hope.' Nothing mattered more to her than winning this race against time; and right then it felt as if time was winning. Nellie sensed her 'gates of hope' closing. 'A wave of despair swept over me,' she lamented. 'What agony of suspense and impatience I suffered that night!' Nellie was on high alert. She said she would circle the globe in seventy-five days, and that is what she was going to do. 'If I fail, I will never return to New York. I would rather go in dead and successful than alive and behind time,' she wrote. Remaining in port for a full day was compulsory under the mail contract. It would be an excruciating twenty-four hours before the SS *Oriental* could depart for Hong Kong. Yet, despite her frustration,

Nellie swallowed her anxiety and stepped ashore. And that is where I meet her.

During my three days in Singapore, I am staying in one of the city's architectural icons, a traditional shophouse in Chinatown. Like books stacked on a shelf, these small businesses, topped with living spaces, stand cheek by jowl on the streets and alleys of the island's historic neighbourhoods. Dating back to the 1840s, Singapore's shophouses caught Nellie's eye when she was here. 'Families seem to occupy the second story, the lower being generally devoted to business purposes,' she wrote. 'Through latticed windows we got occasional glimpses of peeping Chinese women in gay gowns, Chinese babies bundled in shapeless, wadded garments, while down below through widely opened fronts, we could see people pursuing their trades.' Once quarters for a pawn shop and an extended family, the Adler Boutique Hostel now accommodates credit card-carrying backpackers known as flashpackers, prudent business people and independent travellers like me. No traces remain of the pawnbrokers at the Adler now. But, two doors down this stretch of shophouses on South Bridge Road, Heng Fat Pawn Shop still provides short-term loans in exchange for collateral like gold jewellery and branded watches.

Inside the Adler, our dormitory bunk beds are Orient Express-style 'cabins' that come with feather down pillows and duvets, a super-size bath towel, a locker, and a soundproof curtain for privacy. This poshtel (posh + hostel) is Singapore's first and mine as well. I like it. What you forfeit in privacy, through shared dormitories and bathrooms, you make up for in camaraderie with like-minded people. The Adler's glass-fronted lobby exudes the Far East with carved and painted furniture, cushions sewn from indigenous textiles and live tropical plants. It is here that the hostel team helps me map out a trail to trace Nellie's path during her twenty-four hours in Singapore.

Departing from the *Oriental*, Nellie and Dr Brown, a young Welshman, hired a driver and gharry, a 'light wagon with latticed windows … drawn by a pretty spotted Malay pony' from those lined up at Johnston's Pier to take passengers on sightseeing tours. Slatted blinds at the carriage windows shielded Nellie and the doctor from the sun. They drove along a road 'as smooth as a ball-room floor shaded by large trees' into the town. Rules limiting the distance the ponies travelled – no more than ten miles – and the hours they worked – no more than nine – protected the animals from cruelty and exhaustion, Nellie noted. The welfare of the drivers was not mentioned.

The pair's Malay driver and his swift pony, no larger than a rocking horse, took Nellie and the doctor to the Padang, a fifteen-acre lawn facing the bay. The history of the Padang, one of the island's foremost landmarks, stretches almost as far back as the city itself. In land-scarce Singapore where skyscrapers jostle for space, the 'sacred' boundaries of this urban oasis have yet to be encroached. Indeed, with land reclamation, they have been expanded with the addition of Esplanade Park in 1943. The Padang has witnessed many of the country's defeats and triumphs, including the declaration of Singapore's independence on 9 August 1965. Book-ended by the Singapore Cricket Club, circa 1852, and the Singapore Recreation Club, circa 1883, this swathe of green in an ocean of soaring silver high-rises, is a sports hive. Cricketers, footballers and hockey players share the lawn with line dancers and Frisbee flingers. It is also a prime vantage point for viewing the architectural legacy of British rule in a parade of impeccably restored neoclassical buildings. The departure of the British brought new careers for these cloud-white colonial buildings. They have re-invented themselves as swanky cultural venues fringing the Padang: a museum, a theatre, a concert hall and galleries.

The Padang was the first home of the eight-foot high bronze statue of Sir Thomas Stamford Bingley Raffles, known as the 'father of modern Singapore'. Shortly after arriving on the island, it was he who earmarked this land as a recreation ground. Raffles' statue was unveiled in the centre of the Padang on Queen Victoria's Golden Jubilee Day on 27 June 1887. Nellie arrived thirty months later. This is where she would have tipped her deerstalker hat to Sir Stamford. Flying footballs eventually took their toll on the statue. In the interest of preserving both its form and his dignity, Sir Stamford's likeness was relocated nearby in front of the Victoria Memorial Hall on Empress Place during Singapore's centenary celebrations in 1919. There he remains.

When Nellie arrived in Singapore, Raffles was already a legend and this diamond-shaped island ranked as one of the world's major ports. The founder of Singapore has two identical statues dedicated to him –

the original bronze and a replica in faux marble which was erected in 1969, 150 years after Singapore's founding. I visit them both. A three-minute walk from the prototype bronze, the duplicate Raffles stands on the north bank of the Singapore River where this official of the British East India Company first set foot. It is known as Raffles Landing. Given the multi-layered complexity of Sir Stamford, the faux marble concoction of polyurethane resin and marble powder coated in gel might be appropriate for his clone. 'On this historic site Sir Thomas Stamford Raffles first landed in Singapore on 28th January 1819 and with genius and perception changed the destiny of Singapore from an obscure fishing village to a great seaport and modern metropolis', reads the inscription on the lustrous white statue.

Opinions of Raffles shift with time. He has been viewed as the patriarch of modern Singapore, a scholarly expert on the region, a visionary reformer, and alternatively as the brutal face of British imperialism. Whether cast in bronze or faux marble, he exudes an aura of quiet assurance. His stance, tall and erect, arms folded over his chest, right hand casually lifted, left knee slightly turned, is admired by many as valiant; but disdained by some who see it as the power pose of an 'overlord' intent on pursuing the exploitive aspects of colonialism. Is he looking thoughtfully out to sea, or silently asserting his motherland's authority? To be honest, with a frisky September breeze distracting the blazing Asian sun on my first morning in Singapore, I let go of the debates and simply enjoy these serene moments on the riverfront.

Even though his longest tenure here lasted only eight months, the 'father of modern Singapore' is inextricably linked to the island. Wherever you go, there he is. Raffles, or his given name Stamford, is inscribed on roads, shopping complexes, schools, hospitals, a lighthouse and at least two hotels.

The legendary Raffles Hotel maintains its original vocation hosting the wealthy and selling ludicrously over-priced Singapore Slings to the rest of us. Claiming to be 'patronised by nobility, but loved by all', this coconut-white confection is the epitome of colonial Singapore. Tourists like me go to 'drink in' the ambience as much as the gin-based cocktail invented at Raffles in 1915. The Singapore Sling, now the national drink, is a global favourite and a bucket list must. UK celebrity chef Jamie Oliver calls it 'a real posh punch in a long glass, with attitude'. Indeed, it was the Victorian attitude towards women and alcohol that spurred the invention of the Singapore Sling. Raffles' bartender Ngiam Tong Boon created it especially for females more than 100 years ago. Back then women drank only teas and juices, while men knocked back gin and whisky. Even Nellie seemed to comply, content with her cooling lime squashes. This seemingly innocent coral-pink cocktail disguised as fruit juice is actually drenched in gin and liqueurs. Garnished with a pineapple chunk spiked with a maraschino cherry, the Singapore Sling allowed Victorian women to sip away without scrutiny, stigma or shame.

Now it is my turn. It is just before the happy hour reprieve of five o'clock and I am indulging in an authentic Singapore Sling at Raffles' Long Bar. Dozens of other tourists unwind with me amid the bar's earthy decor inspired by 1920s plantation life in Malaya. Vintage palm-leafed fans, *punkah*-style, like the ones Nellie admired at the Grand Oriental in Colombo, wave mechanically in rows to ease the heat. Their silent breezes skim my shoulders. Small burlap sacks packed with roasted monkey nuts sit on the bar and atop adjacent rattan tables, free for drinkers. In a country where littering incurs hefty fines, tossing peanut shells on the floor is positively encouraged here. The bar's brown tessellated floors are carpeted with them. The peanut supply is unlimited; you can eat as many as you like. I do, to dilute the cost of the cocktail. The S$31 (£17.65/$22) charge is actually a small price to pay compared to staying here. A standard single room without breakfast would set me back more than £600 ($750) a night. Back in Chinatown, my bed at the Adler in a female dorm costs the equivalent of £33 ($41) a night, less than the price of two Singapore Slings. Nellie would have had to wait another twenty-six years for the zing – price and taste – of the Singapore Sling. At the time of her visit, Raffles Hotel was emerging from a ten-room bungalow into Singapore's most celebrated lodging, a designated national monument and a world-class icon. Before long, the hotel's lozenge-shaped luggage labels would grace the steamer trunks of the who's who of world travellers. Among them are Joseph Conrad, Albert Einstein, Somerset Maugham, Jean Harlow, Charlie Chaplin, Elizabeth Taylor and Queen Elizabeth II.

Another of Sir Stamford's namesakes, the Raffles Library and Museum, was opened in 1849 to house the colonial leader's extensive natural history collection. Thirty-eight years later the institution moved to its current location on Stamford Road, later becoming the National Museum of Singapore. Nellie and Dr Brown slipped in for a visit here during their gharry tour. A fastidious Nellie deemed the museum 'most interesting'. It is. But I do not have much time to explore the historic collections of this expansive neo-Palladian monument with its celebrated rotunda and dome. I am on a mission to track down a specific Hindu temple that Nellie and Dr Brown tried to visit on their sightseeing tour. Perhaps a museum official can help me to find it.

There are thirty-seven Hindu temples on the island and I am seeking the one that refused admittance to Nellie. I plan to enter it on her behalf, 125 years later. In Nellie's words this 'Hindoo' temple was 'a low stone building, enclosed by a high wall and a gateway swarming with beggars large and small, lame and blind'. As Nellie, Dr Brown and the gharry driver crossed the temple's threshold, they were stopped by Hindu priests. 'My comrades were told that removing their shoes would give them admission but I should be denied that privilege because I was a woman,' Nellie fumed. She was furious. 'Why I demanded, curious to know why my sex in heathen lands should exclude me from a temple, as in America it confines me to the side entrances of hotels and other strange and incommodious things.' The decision was final.

So far nobody has been able to guide me towards the temple – not the team at the Adler Hostel, nor any of the enthusiastic staff at the Singapore Visitor Centres. Maybe this time I will be lucky. Museum press officer Yeo Li Li is dispatched to assist me. I explain my quest, but even this bespectacled senior press officer in Singapore's oldest historical museum cannot identify the temple I am seeking. She sends me off on a temple-hopping expedition to Little India.

Just a fifteen-minute walk northeast from the National Museum, this Indian corner of Singapore is worlds away. Indeed, it is a world of its own. I can hear Little India long before I see it. Bollywood beats blare from enormous speakers in CD shops. Sitar riffs spill from hole-in-the-wall restaurants cooking up tandoori chicken, mutton kebabs, pancake-like spicy dishes called *thosai*, and fish head curries. From tiny basement kitchens, fiery scents ascend and mingle with incense drifting from open-air stalls. Hand-threaded rose, jasmine and frangipani garlands, bright as lanterns, rain down in rows from flower shop ceilings. In the ever-changing Singapore cityscape, Little India refuses to budge. Ducking into temples here, I not only escape the relentless thirty degree afternoon heat, but also the sleek, monotone, charmless side of Singapore. Here people-scale buildings are splashed in eye-popping blues, pinks, greens and yellows. Curry houses thrive alongside silk stalls, bangle shops, sari specialists, fruit stands, and a divine collection of Hindu temples.

One of Singapore's oldest temples, and therefore the top contender in my search, is Sri Srinivasa Perumal which is dedicated to Lord Vishnu, the preserver and protector of the universe. It sits just off Serangoon Road, Little India's main thoroughfare. I remove my sandals at the entrance and proceed under the Wedgewood blue wooden arch – the temple's monumental gateway known as a *gopuram*. Above me, five tiers of life-size, lifelike ceramic human and animal figures ascend into the sky like a pyramid. They are incarnations, avatars of Vishnu and other Hindu deities. Some are seated, but most stand with one forearm raised and two fingers pointing towards heaven as they gaze out over the Singapore horizon. Inside, except for a quartet of pigeons, I have the prayer hall to myself. Green meadow-like ceilings speckled with daisies host enormous swirling *mandalas*, the ornamental circles that symbolise the universe and enlightenment. Along the prayer hall corridors, I encounter treasures such as a five-headed cobra throne, a gold-encrusted chariot, flying gods and multi-armed goddesses. But I have not found the temple that turned Nellie away.

I never will. My search is in vain. The temple that incensed Nellie, called Sri Sivan, is not in Little India. It never was. A hostage to Singapore's supersonic ascent from colonial port to a world-class city-state-nation, Sri Sevan Temple was moved and rebuilt no less than three times. When Nellie was here, the temple sat about a mile south of Little India on Orchard Road, on land purchased from the East India Company. The area was known as Dhoby Ghaut, then the realm of the dhobies, laundrymen from southern India who served the British military establishing the settlement of Singapore. From sunrise to sunset, dhobies could be seen knee-deep in the nearby Stamford Canal pounding clean the linens of their colonial patrons. Sri Sevan Temple was where they worshipped. Today, all that survives of their legacy is the name. It has been allocated to Dhoby Ghaut MRT station, a hub for three major metro lines, and Dhoby Ghaut Green, a patch nearby where the dhobies may have spread their laundry to dry. On the site where the laundrymen once prayed, hundreds of thousands of people now pass through on pilgrimages to multi-story shopping meccas strung along Orchard Road. When Nellie was here, Orchard Road was a shady lane bounded by nutmeg plantations and the fruit orchards that provided its name. Nellie noticed the absence of pavements. Now they surge like superhighways for more than a mile along Asia's most famous shopping street.

Just a few steps along these pavements, not far from the site of the Sri Sevan Temple that excluded Nellie and eluded me, is the former colonial governor's house. Nellie and Dr Brown were received there by then Governor Sir Cecil Clementi Smith. Now it is me that is denied entry. I cannot even sneak a peek at the Palladian mansion where the pair were entertained. Now it is used by Singapore's first female President Halimah Yacob, to host official guests. I am not official. Called the Istana, which means 'palace' in Malay, the site is only open five days a year. This is not one of them. Two Royal Ceremonial Guards stand erect and rigid on sentry duty, one on each side of Istana's hefty white wrought iron gates. The guards are dressed in immaculate white long-sleeved jackets with epaulettes; a red stripe runs vertically from waist to ankle on their dark trousers. A police officer, arms crossed with authority over his chest, walkie-talkie in hand, reinforces the defence. There seems no way I can snatch a photograph through these forbidding gates. A menacing 'no' is the swift answer when I wander up to ask, explaining that I have travelled from London to retrace the steps of Nellie Bly who was welcomed here by the governor 125 years ago. Twenty years before that, in 1869, this governor's house was constructed on the site of a former nutmeg plantation by the convict labourers from India, Ceylon and Hong Kong who built early Singapore. I stand where Nellie once stood, but the gates that were open to her are closed to me.

After paying their respects to the governor, Nellie and Dr Brown enjoyed an early dinner on the veranda at the Hotel de l'Europe. She described the hotel as a 'long, low, white building set back in a wide, green lawn, with a beautiful esplanade, faced by the sea, fronting it'. They sat at extended white tables 'where a fine dinner was served by Chinamen'.

Now just a memory, the Hotel de l'Europe once vied with Raffles for the custom of colonials and tourists. *The Jungle Book* author Rudyard Kipling, who was touring Asia eight months before Nellie, preferred to dine at Raffles, but chose French-run Hotel de l'Europe for his lodgings. Nellie and Dr Brown had no need to stay; the time had come to return to the *Oriental* for boarding.

The two were on their way back to the port when their gharry driver made a brief, but fateful stop at what Nellie called his 'humble home'. A monkey stood near the front door, his long, tubular tail looped around him. Inside, the driver's 'pretty little Malay wife', wrapped in a linen sheet, was looking after her babies. She had a 'large gold ring in her nose, rings on her toes and several round the rims of her ears, and gold ornaments on her ankles,' Nellie noted. But it was not the driver's exotic wife or their infants that captivated her. It was the macaque monkey outside. She could no longer resist. 'When I saw the monkey my willpower melted and I began straight away to bargain for it,' she wrote. Nellie may have taken into account that she was already half-way around the world and well on her way to San Francisco. Or she may simply have been impetuous. In any case, she bought the monkey.

Nellie paid $3, the equivalent of about £69/$86 today. Her purchase that evening would end up costing much more than money. The monkey would cause nothing but trouble for the rest of Nellie's global journey, and even afterwards. The gharry driver had assured Nellie that the 'monkey no bite'. But McGinty – as he was ultimately named after trying out others such as Solaris, Tajmahal and Jocko – certainly did bite. The steamship crew who attended to the monkey would bear the brunt of his anger, emerging from below deck with bites and scratches. Later in Hong Kong, when McGinty was transferred from the SS *Oriental* to the SS *Oceanic* for the voyage to Japan, Nellie noticed a stewardess whose arm was bandaged from her wrist to her shoulder. 'What did you do?' Nellie asked in consternation. 'I did nothing but scream; the monkey did the rest!' the stewardess replied. McGinty did not like his mistress either. One day aboard the *Oceanic*, Nellie visited him only to discover that young men had been plying him with alcohol. 'It was holding its aching head when I went in, and evidently thinking I was the cause of the swelling, it sprang at me, making me seek safety in flight,' she wrote.

Despite it all, Nellie would remain loyal to McGinty, even saving his skin when the crew on the *Oceanic* threatened to throw him overboard. After several days of brutal winds and fierce rains as they crossed the Pacific, the sailors told Nellie that McGinty had to go. Monkeys were 'Jonahs' that brought bad luck and danger to ships, they said, blaming McGinty for the barbarous storms that would not cease until they disposed of him. In Nellie's words: 'A little struggle between superstition and a feeling of justice for the monkey followed.' When she discovered that ministers were also thought to be 'Jonahs', Nellie agreed that McGinty could be flung off the ship if the two churchmen on the *Oceanic* were also sent overboard. 'Thus the monkey's life was saved,' she wrote. But his close call with death did not improve McGinty's affection for Nellie ... or his conduct. When they returned to New York, the monkey is said to have wrecked her apartment. Even so, McGinty became one of the icons associated with Nellie's journey along with her ever-present deerstalker cap, broadcloth dress and ulster coat. He is as much a part of her voyage as the small leather gripsack Nellie carried around the world.

Chapter 8

In Which Nellie Experiences Peaks and Troughs

Peak Tram, Hong Kong

Hong Kong
23–24 December, 26–28 December 1889

'I stared at him; I turned to the doctor; I wondered if I was awake; I concluded the man was quite mad, so I forced myself to laugh in an unconcerned manner, but I was only able to say stupidly: "The other woman?"'

Nellie Bly

Nellie sailed into the colonial port of Hong Kong in the aftermath of a raging monsoon that nearly crushed her. I flew in amid escalating typhoon warnings. The SS *Oriental* delivered her passengers to Hong Kong at daybreak. My flight arrives into Hong Kong International Airport in the late afternoon. Nellie's tempest was over; mine was about to begin.

The World's globetrotter was ready to leave Hong Kong as soon as she arrived. Nellie had a race against time to win. Even so she could not overlook the splendour of the scene that greeted her as the SS *Oriental* steamed into Victoria Harbour on the sunlit morning of 23 December. The harbour bay was a 'magnificent basin, walled on every side by high mountains,' she wrote. The blazing orange sun had transformed the waters into an enormous mirror that reflected the green-ribbed mountains soaring into a cloudless sky. A parade of gleaming white houses ascended tier after tier up the mountainside like a 'gigantic staircase of fairy tale castles,' Nellie wrote. 'Hong Kong is strangely picturesque,' she said of the terraced colonial city. Alongside the *Oriental*, she noticed 'strange craft from many countries'. Anchored among the native Chinese junks and sampans, the bay was peppered with warships known as ironclads, torpedo boats, mail steamers and Portuguese lorchas with their batten sails.

But Nellie's main focus was the wooden pier nearby where she would hire a sedan chair to carry her from the port into the city. 'My only wish and desire was to get as speedily as possible to the office of the Oriental and Occidental Steamship Company to learn the earliest possible time I could leave for Japan,' wrote Nellie. In haste, she disembarked the SS *Oriental* accompanied by her Welsh friend Dr Brown. They mounted sedan chairs and set off of at a 'monotonous trot' that reminded Nellie of a 'pacing saddle horse'. Their carriers, with 'untidy pigtails twisted around their half-shaven heads', loped along the shore and up a route tracing the mountain. The pair were headed along the oldest road in Hong Kong, straight to the offices of the steamship company's agent C. D. Harmon at 50A Queen's Road Central.

Nellie was on cloud nine. Day thirty-nine of her global journey and already she was in China, British China. 'I was leaving particularly elated, because the good ship *Oriental* not only made up the five days I had lost in Colombo, but reached Hong Kong two days before I was due.' On this, her maiden voyage to China, the *Oriental* had broken all previous records for speed between Colombo and Hong Kong. 'I went into the O. and O. office feeling very much elated over my good fortune,' she wrote, 'with never a doubt but that it would continue.' Luck was with her for the long haul … or so Nellie thought.

Within moments of arriving at the agent's office, Nellie's joy dive-bombed to despair. Her world race was in jeopardy, the agent, Mr Harmon, informed her.

'Aren't you having a race around the world?' he asked me, as if he thought I was not Nellie Bly.

'Yes; quite right. I am running a race with time,' I replied. 'Time?' he said 'I don't think that's her name.' 'Her! Her!!' I repeated, thinking, 'Poor fellow, he is quite unbalanced, and wondering if I dared wink at the doctor to suggest to him the advisability of our making good our escape.'

'Yes, the other woman; she is going to win,' [Harmon told her.] I stared at him; I turned to the doctor; I wondered if I was awake; I concluded the man was quite mad, so I forced myself to laugh in an unconcerned manner, but I was only able to say stupidly: 'The other woman?' 'Yes, the other woman; she is going to win,' the agent said. 'She left here three days ago.'

Unbeknownst to Nellie until that very moment, she had a competitor. A competitor by the name of Elizabeth Bisland. And Elizabeth Bisland was in the lead. Nellie was incredulous. Her lips quivered and her eyes grew wide. For several moments she could not breathe. But it was true, and it stung. Elizabeth Bisland, a journalist sent by *The Cosmopolitan* (the forerunner of today's best-selling *Cosmo* magazine) had set out from New York the day Nellie departed, circling the world in the opposite direction. Nellie was travelling east as Elizabeth travelled west. Elizabeth had departed from Hong Kong on P&O's SS *Thames* seventy-two hours earlier. The two women's ships, and their destinies, had likely crossed in the South China Sea.

Nellie stood rigid. It took all of her inner strength to maintain her composure as the steamship agent spewed out such unbearable news. Harmon did not stop there. The wound deepened … and then he poured salt in it. She would not be going anywhere anytime soon. Nellie would be stranded in Hong Kong for five excruciating days until the SS *Oceanic* arrived. She would lose 127 hours and twenty minutes in British China, as recorded in her logbook. 'That is rather hard, isn't it?' Nellie said in true Bly-style, forcing a fake smile 'that was on the lips, but came from nowhere near the heart'. Mustering up every ounce of dignity, and bravado, she set the record straight. She was not racing against anyone. Time was her only challenge. 'I promised to do the trip in 75 days and I will do it.' But she did not. Nellie completed the journey in seventy-two days; Elizabeth Bisland finished in seventy-six days. The winner took all. The rest is history.

*   *   *

With five eternal days to wait until she could board the SS *Oriental* for Japan, Nellie had more than enough time to explore Hong Kong. Tucked in between mountains and sea, even then this colonial island was packed; 1,600 people in the space of an acre, according to Nellie. 'They remind me of a crowd of ants on a lump of sugar,' she wrote. Today Hong Kong has the fourth highest population density on the planet after Macao, Monaco and Singapore.

In Nellie's era, human-powered rickshaws and sedan chairs traversed earthen roads framed by arcades strung with gaudy banners that hung like drapes above shops and restaurants. Peddlers hauled their wares in baskets slung from bamboo poles, like scales, across their thin shoulders. Conical hats, parasols and awnings, woven from straw and reeds, curtailed the sun. Most of the mountainsides were visible and buildings reached no higher than six storeys. Queen's Road Central, site of the O&O agent's office, was the main thoroughfare. Named after Queen Victoria, it was renowned for shopping even in Nellie's day. 'Interesting to all visitors,' wrote Nellie. Best of all were the exotic Chinese emporiums lined with treasure-filled black wooden cabinets. 'One feels a little thrill of pleasure at the sight of the gold, the silver, ivory carvings, exquisite fans, painted scrolls,' wrote Nellie, 'and the odor of the lovely sandal-wood boxes, coming faintly to the visitor, creates a feeling of greed,' she admitted. 'One wants them all, everything.'

Today on Queens Road Central Nellie would encounter global luxury brands such as Gucci, Prada and Louis Vuitton in chic showrooms stylishly stacked in atriums and malls. In this international retail melting pot, she would discover the same ubiquitous fashion brands and chains – Gap, Topshop, HMV, Starbucks – found in cities around the world. With few exceptions, Nellie would see that yesterday's indigenous commerce has been obliterated by today's globalisation.

Hong Kong's cuisine has fared similarly. Restaurants offering plant-based burgers, craft beers and bagels, along with dishes from European celebrity chefs, cohort with dim sum tea houses and street food stalls where curry fish balls simmer in large black skillets. Nellie mentioned nothing about the elegant colonial restaurants where she would have dined while on her 'enforced stay' in Hong Kong, but in her book she vividly describes a local canteen. 'I came upon an eating house, from which a conglomeration of odors strolled out and down the road. Built around a table in the middle of the room, was a circular bench,' she wrote. 'The diners perched on this bench like chickens on a fence, not letting their feet touch the floor … sitting down with their knees drawn up until knees and chin met; they held large bowls against their chins, pushing the rice energetically with their chop-sticks into their mouths.' Today's counterparts are the street food stands where spicy aromas escape from fizzing woks, revolving spits and portable grills. Stainless steel chafing dishes and platters are heaped with rice noodle rolls, stinky tofu, grilled squid tentacles, roasted chestnuts and egg tarts. There are no benches, no tables and no chairs. There is not enough room.

Space has always been at a premium in Hong Kong. In this Special Administrative Region of China, skyscrapers rise suddenly out of nowhere and businesses open and close in the blink of an eye. With space so tight, heritage is rarely indulged, leaving Hong Kong largely bereft of evidence from its past. Unlike cities with visible history, tangible vestiges of former times have vanished here. They endure mainly in photographs, memoirs and a smattering of sites that have survived the skyscraper invasion. Finding Nellie here is hard for me; what is left of her path is scattered. With a typhoon on the warpath, I have only a day to find what remains of the late 1880s when she was here. Now I am the one who is racing, racing to outpace the rage of Typhoon Kalmaegi. My Elizabeth Bisland is a typhoon. This will require a crack of dawn departure from my hotel in Hong Kong's mid-levels just above the downtown business district.

It is 6.00 am and I am getting a head start on the day and the Hong Kong rush hour, but most of all the approach of the typhoon now coming my way at ninety-eight miles per hour. I activate my city survival skills, the steely assertiveness that I have honed in London. But my 'Square Mile' savvy is no match for the throngs of Hong Kong.

My first destination, Happy Valley, is seven stops and twenty minutes away by tram. I drive myself like a car through scores of scurrying commuters that I was hoping to avoid at this early hour. We are all heading for the Happy Valley tram. Nellie, escorted by the SS *Oceanic*'s Captain Smith, arrived in Happy Valley in *a jinrikisha* on Christmas Eve morning, 1889. 'Where the mountains make a nest of one level, green space', is how Nellie described it. The only flat ground in steep-sloped Hong Kong, Happy Valley's former malaria-ridden paddy fields were cleared by the British for a racecourse in 1846. Today it stands as a grassy oasis, a vast amphitheatre where race fans flock on Wednesday evenings from July to September to place their bets at the only venue where gambling is legal. Up to 55,000 locals and tourists pack into the seven-storey grandstands that have long since replaced those constructed from palm trees when Nellie was here.

Today, Hong Kong's tallest residential towers sprout like concrete forests around this former bog turned racecourse. Two of these soaring silver pinnacles – Highcliff and the Summit – are so tall and so thin and so close they are known as the Chopsticks. Far below the skyscrapers, and just above the racecourse, six multi-faith cemeteries stand frozen in time. Consecrated in the 1840s, they seem oblivious to the urban jungle surrounding them. The Colonial Cemetery, the Jewish Cemetery, the Hindu Cemetery, the Parsee Cemetery, the Catholic Cemetery and the Muslim Cemetery; together they create a stadium of tombstones terracing Happy Valley's foothills like rotten teeth. Headstones point heavenwards in perfect symmetry with the skyscrapers subsuming them. It is as if Happy Valley's sky dwellers are vying for space with 10,000 corpses asleep in their graves below. This urban nightmare was once a dream. Nellie raved about Happy Valley. 'The Fire Worshippers [Zoroastrians] lie in ground adjoining the Presbyterians, the Episcopalians, the Methodists and the Catholics, and the Mahommedans are just as close by,' wrote Nellie. 'That those of different faiths should consent to place their dead together in this lovely tropical valley is enough to give it the name of Happy Valley, if its beauty did not do as much,' she wrote with passion. 'One wanders along the walks never heeding that they are in the Valley of Death, so thoroughly is it robbed of all that is horrible about graveyards.'

Happy Valley was like heaven to Nellie. 'It rivals in beauty the public gardens and visitors use it as a park,' she wrote. She was describing the archetypal garden cemetery, sometimes called rural cemetery, that was Happy Valley – acres of landscaped countryside with floral borders, ornamental ponds and garden paths for strolling. Garden cemeteries were the horticultural wonders of their day. They offered a solution to the overcrowded urban cemeteries of the nineteenth century – and the stench, disease and rats they spawned. Overflowing cemeteries meant that bodies were often buried within inches of each other. Coffins leaked and epidemics escalated before city cemeteries were sent into surrounding rural areas to become gardens of remembrance, the forerunner to public parks. Père Lachaise Cemetery in Paris led the movement towards these 'gardens of graves' when it opened in 1804. This French graveyard, now the most visited necropolis in the world, was the inspiration for garden cemeteries across Europe, America and colonial Asia.

Happy Valley is sad now. I am too as I wander through these forlorn cemeteries longing for Nellie's descriptions to return to life. This City of the Dead – once alive with flowing fountains, flower beds, vine-draped arbours and pedestal bird baths – is lifeless, deserted and dangerous. Except for a few historians and relatives paying tribute to their ancestors, hardly anybody visits the Happy Valley cemeteries anymore. Signs warn visitors of slippery ground and disintegrating stairs, and, worst of all – snakes! This morning I am the only living soul in a massive, crumbling, reptile-infested cemetery, stepping warily (and loudly to ward off any snakes), across broken concrete and overgrown paths to explore the many faith graveyards.

Bereft of the beauty Nellie described, a sinister eeriness now shrouds the cemeteries – drunken tombstones cloaked in black moss, fallen angels nursing shattered wings, granite-lipped virgins bearing crumbling crosses, fractured urns and ghost-finger ferns. For those with no fear of the supernatural, they offer an evocative window into Hong Kong's past. So much history lies asleep here, so many remarkable lives. But I must pull myself away and leave them to rest in peace. Typhoon signal levels are rising, along with a torrid mid-morning sun. I exit through the Muslim burial ground, and pause to read the cemetery rules: 'Visiting graves is encouraged in Islam which benefits both the dead and the living.

While it is a tribute paid to the dead, it prompts the living to think about the essence of life.'

It is something to contemplate as I flee the hottest site in Hong Kong – low-lying (un)Happy Valley for the coolest – Victoria Peak, with a stop for the legendary Noonday Gun salute on the waterfront at Causeway Bay. It is 11.30 am. The orphic lure of the Happy Valley cemeteries has detained me. Now is my only chance to experience the marvellously eccentric tradition of the Noonday Gun. I weave through pedestrians, in the now sweltering heat, to hop the tram for the four stops and seven-minute dash to the Noonday Gun that waits for no one. I arrive with a just few moments to spare, and join the small flock of tourists gathered for the daily blast celebrated in Noël Coward's song *Mad Dogs and Englishmen* [Go Out in the Noonday Sun]. In this piercing midday sun, spouting beads of sweat across my forehead, I am an Englishwoman feeling hot and bothered, but relieved to have arrived in time. Coward's 1930s lyrics, a clever blend of complicated rhythms and rhymes, sound in my head in his clipped and articulated tones. Reference to the Noonday Gun arrives in the penultimate verse: *In Hong Kong, they strike a gong, and fire off a Noonday Gun*. Coward himself fired off the Noonday Gun in 1968. The playwright and composer is said to have completed the entire ten verses while driving between Hanoi and Saigon 'without even the aid of pencil and paper'. 'I sang it triumphantly and unaccompanied to my travelling companion on the veranda of a small jungle guest house,' Coward said, noting that the gecko lizards and tree frogs gave every vocal indication of enthusiasm'.[1]

The Noonday Gun tradition originated in 1864, sixty-seven years before Coward penned his witty ditty, and twenty-five years before Nellie visited Hong Kong. For unconfirmed reasons, the Royal Navy ordered Noonday Gun proprietors, multi-national giant Jardine Matheson Holdings Limited, to fire a one-shot salute daily at noon in perpetuity. Jardines continues to honour the pact more than a century and a half later, now raising money for charity by 'selling' the right to perform the legendary ritual for a substantial (more than £3000) donation to Hong Kong's Community Chest.

Today a Jardines guard, clad all in black except for snow-white gloves, will perform the ritual. I move closer to admire the quick-firing three-pound Hotchkiss naval gun – a slick, but anorexic, cousin of a cannon – its polished brass and Pacific blue streamlined body mounted on a pedestal and roped off with golden chains. Solemnly, the guard marches from his station to the gun's elevated enclosure, unlatches the chains, feeds it a cylindrical shell, aims it across the bay and cocks it ready. He returns to his station, checks his watch, strides to a brass bell suspended from a nearby pole, chimes it twice and twice again. He marches to the gun, seizes a lanyard, and yanks it hard with his right hand while his left hand flies up in a swift salute. The explosion fractures the air as it ricochets across Causeway Bay. Clouds of swirling smoke obscure the gun. The guard performs a re-run of the double bell chimes – and the 155-year-old ritual is complete. It is 12.02.38 pm and we have experienced the timeless blast of the Noonday Gun. In Nellie's day it was a more ear-splitting explosion. Jardines reduced the power, and the decibels, in 1961.

The next stop on my typhoon-fuelled trail transports me from just above sea level at Causeway Bay to Hong Kong island's highest peak at 1,818 feet. I am riding the historic Peak Tram to Victoria Gap exactly like Nellie did when she was here. Asia's first funicular, and for at least a century, the steepest on the planet, the Peak Tram is a stunning feat of Victorian engineering. Nellie boarded the tram seven months after it was launched in May 1888. Scooting nearly a mile above the city's central district to Victoria Gap, known as the Peak, the tram was the only way up, except by sedan chair, until 1924 when the Old Peak Road opened. At eight minutes from bottom to top, the tram journey remains the speediest traverse. Buses take as much as an hour to circle up to the peak. When Nellie was here she paid 30 American cents up and 15 cents down for a first class seat on the steam-powered tram. Today the tram runs on electricity and it is a £5.50/$7 round trip wherever you sit. During its first year, the tram transported 150,000 passengers up to Victoria Peak. Today, the tram transports more than six million passengers a year.

I join the other passengers squeezing through the narrow doors to board the tram. A chorus of anticipatory chatter accompanies us. Our seats, slatted wooden benches, face uphill so we will not fall forward. I put myself in Nellie's 'place', which is not difficult inside this tram which seems little changed since her visit. Outside is a totally different story. Like an aeroplane lifting off, the tram glides up and out of its gloomy terminus and emerges into the shimmering afternoon daylight. Tree branches are close enough to touch. Within minutes Hong Kong's spindly skyscrapers begin to swallow the skyline, slanting forward like keyboard slashes. At least that is the way they appear from this dramatic forty-five-degree incline. Nellie would have seen only trees. At times we veer close enough to these sky residences to catch glimpses inside. Picture a jolly, candy-red Victorian tram gliding past futuristic skyscrapers. Thomas the Tank Engine chugging through Las Vegas would be just as absurd.

As we mount, the rhythmic whir of the tram is punctuated by metallic clanking, like a roller coaster in slow motion. Except for the tell-tale skyscrapers, I can almost imagine Nellie sitting beside me, her skirts gathered round her feet, her gripsack planted on her lap. That illusion is immediately dashed as the tram arrives at its destination and we are invited to alight on the right for the Peak Tower. Stepping away from Nellie's world, I am fast-forwarded into a 'multi-level shopper's paradise' that is today's 'Peak experience'. Shopping literally reaches new heights as consumers swoon over Swarovski crystals, Swatch watches, designer sunglasses and Chinese souvenirs. It looks like a stunted champagne glass to me, but the Peak Tower is lovingly called 'the rice bowl' or 'the wok' by Hongkongers. I want to jump back on the tram and hide in its faux Victorian reality.

Outside this shopping 'wok', the pre-typhoon, post noonday sun is ruthless, even up here at the Peak where folks have come for centuries to escape Hong Kong's relentless heat and humidity. I have come for the view. Like a twenty-first century Emerald City of Oz, the soaring superstructures of Hong Kong thrust out of the sea in glistening pinnacles straight up towards the Peak. It is hard to shift from this exhilarating view, but I must. Torrential rains are forecast. The typhoon signal level is one now; once it reaches level eight Hong Kong will batten down its hatches. Everything shuts. The pressure is on. I still have a Nellie-related quest to fulfil here. Once she arrived in the area where I stand, Nellie was carried by sedan chair to the summit, Hong Kong's highest point. I want to follow her trail and track down the 'umbrella seat' where she and her three bearers rested on their way up. 'At the Umbrella Seat, merely a bench with a peaked roof, everybody stops long enough to allow coolies to rest,' wrote Nellie, 'then we continue on our way, passing sight-seers, and nurses with children. After a while they stop again, and we travel on foot to the signal station.'

Pre-visit research provided no clues about the location of the umbrella seat and even whether it still exists. I am hoping that someone in the Peak Visitors Centre, housed in a 1956 tram car, can help. But centre staff Sanford Lee and Windie Chui have never heard of Nellie's umbrella seat. As has happened before on the Nellie Bly trail, my mission has now become theirs. Fingers fly like seagulls across tablet screens as they seek the elusive bench in cyberspace. Nothing is revealed. Then Sanford, a youthful looking senior who has served tourists to the Peak for decades, has a hunch. 'I think I might know what you are looking for,' he announces. His second hunch – that I might never find it on my own – means that he sends Windie with me. Shattered by the searing heat, it is only my pride that keeps me on pace with the young and exuberant Windie as we climb up, and up, and up the peak. Winding her shoulder-length black hair into a ponytail, Windie is also feeling the heat. She carries a tablet for on-the-go enquiries. A mobile phone to contact Sanford is tucked in a pocket of her black jeans. Eventually tourist traps surrender to gardens, then grassy hillsides. There it is, twenty minutes later, settled on the slopes, an elegant stone structure with a canopy that is shading a bench just right for a rest stop. Windie and I politely dislodge a pensioner occupying our umbrella seat and take turns posing before it. She seems just as excited as me to have located the seat. Now that we have achieved our quest, I am ready to flee the sun and beat the typhoon. Not Windie. We head another fifteen steamy minutes up towards the summit and the site of the former signal station viewed by Nellie, now a jungle of radio towers. Signal flags were used to announce the arrival of ships, incoming post and approaching weather at Victoria Peak Weather and Signal Station before the onset of telegrams, telephones and the internet. We are at the top; the ultimate panorama of Hong Kong fans out below us. Nellie called this view 'superb' and describes it in the most romantic terms.

'The bay, in a breastwork of mountains, lies calm and serene, dotted with hundreds of ships that seem like tiny toys. The palatial white houses come half-way up the mountain side, beginning at the edge of the glassy bay. Every house we notice has a tennis-court blasted out of the mountain side.' Like me, Nellie was not here at night, but she had heard of its splendour. 'They say that after night the view from the peak is unsurpassed. One seems to be suspended between two heavens. Every one of the several thousand boats and sampans carries a light after dark,' she wrote. 'This, with the lights on the roads and in the houses, seems to be a sky more filled with stars than the one above.'

Today Nellie's stars are concealed by high-rises. Hong Kong boasts more 'official' skyscrapers than any other place on the planet; 303 buildings with a minimum height of 330 feet high. New York City is in second place, with 237 skyscrapers. Nellie's global route today includes six of the ten cities with the largest number of skyscrapers: Hong Kong, New York, Chicago, Tokyo, Guangzhou (Canton) and Singapore.

*   *   *

The rain is falling in sheets by the time Windie and I trudge back down to the Peak Visitor Centre. The deluge obscures everything, even the massive Peak Tower. The typhoon signal has jumped from one (standby) to three (strong winds). I say my thank yous to Windie and Sanford and sprint back to the Peak Tram to descend to a once-heaving city now in shut-down mode. It is only 6.00 pm, but Hong Kong is dark and eerily empty. Crossing the roads and concrete corridors that link the levels of this metropolis, I arrive back at my hotel, the Bishop Lei International House, to discover that typhoon signal eight (gale and storm force winds) is now in effect. The hatches have been battened. I have won my own race against time today... just. But can I make it to Canton, the next stop on Nellie's path, tomorrow?

*   *   *

Frustrated and bored by her imposed stay in Hong Kong, Nellie decided to visit the exotic port city of Canton, today known as Guangzhou. She felt she had already seen 'everything of interest' in Hong Kong and needed a distraction. Canton, ninety miles up the Pearl River, was within reach. It was an authentic Chinese city. Nellie wanted to see Chinese people while she could. Back at home, the United States government had blocked their entry with the Chinese Exclusion Act of 1872. The former universal melting pot was transforming into a sieve. 'I knew we were trying to keep the Chinese out of America so I decided to see all of them I could while in their land. Pay them a farewell visit as it were,' she wrote.

She booked a passage on the night sailing of the SS *Powan*, a Pearl River steamer in the fleet of the Hong Kong, Canton & Macao Steamboat Company. It would deliver her to Canton the next morning. Nellie would spend Christmas day in Canton and return to Hong Kong that night. With American Captain Grogan in command, the *Powan* cast off on Christmas Eve. The captain was the 'fattest and most comically proportioned' man Nellie had ever seen. She had to stifle her laughter at his 'roly-poly body, his round face embedded, as it were, in the fat of his shoulders and breast'. He did not seem to have a neck. Nellie knew first-hand how her giggles would insult the amiable captain. She recalled her own sensitivity when remarks were made about her own appearance – 'the shape of my chin, or the cut of my nose or the size of my mouth'. Her advice to detractors: 'criticize the style of my hat or my gown, I can change them, but spare my nose, it was born on me'. Following her own counsel, Nellie wrote: 'Remembering how nonsensical it is to blame or criticise people for what they are powerless to change, I pocketed my merriment, letting a kindly feeling of sympathy take its place.'

Soon after the *Powan*'s departure 'everything was buried in darkness,' wrote Nellie. Surrendering to the 'infinite radiance' of the stars above and the silky black river below, she let go of day forty-one, releasing herself from the dogged pursuit of the race. Lulled by the gentle lapping of the mighty Pearl, the third longest river in China, Nellie fell into a trance as the *Powan* 'softly and steadily swam on'. It was a rare moment of abandon on Nellie's worldwide journey.

'To sit on a quiet deck, to have a star-lit sky the only light above or about, to hear the water kissing the prow of the ship, is, to me, paradise,' she wrote. 'They can talk of the companionship of men, the splendor of the sun, the softness of moonlight, the beauty of music, but give me a willow chair on a quiet deck, the world with its worries and noise and prejudices lost in distance, the glare of the sun, the cold light of the moon blotted out by the dense blackness of night. Let me rest rocked gently by the rolling sea, in a nest of velvety darkness, my only light the soft twinkling of the myriads of stars in the quiet sky above; my music, the round of the kissing waters, cooling the brain and easing the pulse; my companionship, dreaming my own dreams. Give me that and I have happiness in its perfection.' But she did not indulge herself for long. Reality swept in and shattered the spell, casting Nellie from her 'nest of velvety darkness' and bombarding her perfect happiness with the challenge she faced. 'But away with dreams,' she wrote. 'This is a work-a-day world and I am racing Time around it.'

*   *   *

Relentless rain lashes the floor-to-ceiling window in my twenty-first storey hotel room at Bishop Lei International House. Spiralling out of control all night long, the typhoon is pummelling the area. I am tracking the typhoon's 'progress' on the Hong Kong Observatory website, hoping against hope that the winds will die down. Satellite images showing a gargantuan tornado – a frenzied whirlpool of clouds – offer no comfort. Flights are cancelled, schools and businesses are closed, even the stock market has ceased to trade – on the very morning that I am due to depart for Canton. Typhoon, or no typhoon, I must get there. It was one of the most exotic stops on Nellie's world tour and the place where she spent Christmas day 1889. Besides, I have already invested time, money and mountains of stress to get a visa for mainland China. It does not help that I chair a London-based organisation working to advance human rights in China. I kept it quiet, but I think the officials know. They seem to know everything. My journey is about Nellie Bly, so I decide to adopt her steel rod single-mindedness and separate the 'business' of China's brutal disregard for basic human rights from my quest to follow in her footsteps. I certainly cannot risk jeopardising the work of my charity and our courageous Chinese partners, human rights defenders who endure torture and imprisonment in attempts to hold their government to account. In any case, I will be homeless in Hong Kong because the Bishop Lei International House run by the Catholic Diocese of Hong Kong does not have a room for me tonight. The front desk staff could be robots; their actions are methodical, their faces expressionless. They take no account of my situation. I am on my own. There is nobody to consult. So what do I do now? Try to find somewhere to shelter and spend another night in a zombie-like metropolis? Place myself at the mercy of Kalmaegi's almighty gusts to battle my way to Hong's Kong's Hung Hom station. Even if I can get there, will the station be open? Will trains be running to Canton? If planes are grounded, will trains be suspended? And finally, I ask myself, what on Earth would Nellie Bly do?

That is the one question I can answer. Out comes the resolve; on comes the waterproof. Abandoning Bishop Lei's dry but inhospitable shelter, it is now me versus Kalmaegi. Peering through the blustery deluge, the rain-slick streets are deserted as far as I can see. The wipers have stopped working on this massive urban windscreen. Any other time, Hung Hom station is minutes away by metro or bus. Now Mars is closer. Fierce winds snap the rain away and save me from a drenching, but they demand full force balance in return. It is just me and Mother Nature.

But not for too long. It cannot be. It is not possible. I squint, cup my hands against my eyebrows in a ten-finger shield. It is! It is! In the distance I can make out a lone vehicle defying the tempest. My fantasy taxi has materialised; come to life! Dashing into the road, I wave down the valiant driver. If he is up for challenging Kalmaegi, so am I. Silently, we crawl along Route One across Victoria Bay. I do not want to distract him with conversation; but I am spurring him on with unspoken encouragement. I shut my eyes and breathe deeply, slowly. The tension tastes like steel in my mouth. Three eternal miles later, I land on Mars, to a deserted, but functioning train station. Spouting effusive thank yous to the heroic driver, I grab my bag and dash inside.

Typhoon Kalmaegi has paved the way for my rail journey to Canton. A perpetually heaving Hung Hom station is all mine. I cross a vast and silent concourse searching for the ticket office. Nobody is queuing. I am the only passenger. It is surreal. I sail through security, customs and immigration into a dark and vacant departures hall where there is no need to linger. The Hong Kong–Guangzhou Intercity Express awaits me on platform five. I board for the two-hour journey to Nellie's Canton and my Guangzhou.

Royal blue recliner seats, arranged in pairs on both sides of a long corridor, await their Tuesday morning business passengers; but there is just me. Snow-white headrest protectors, like collars, are positioned on row upon row of unoccupied seats. After an endless day and a sleepless night, this solitude is just what I need to de-typhoonise. My ticket reservation leads me to carriage seven, seat thirty-two on this virtual ghost train. I make my way along the corridor only to find that seat thirty-three, next to the window, is occupied. I have the aisle seat. Why are the only two passengers in a carriage with seventy-two seats placed right next to each other?

My seatmate grins at me. He is a rather dapper young Asian who has also outfoxed Kalmaegi. We share our names and then our 'war stories'. Adam walked right through the typhoon to get to Hong Hum station. I tell him about my phantom taxi driver. Adam is a Cathay Pacific flight attendant from Malaysia. In between trips, he is making a quick train journey to Canton to visit a friend. His short-cropped porcupine hair crowns a broad forehead above deep-set eyes and a kind smile. He is a thirty-something who looks no older than 21.

We must be in first class because a waiter arrives to take our orders for breakfast. It is a victory breakfast for both of us after tangling with a tireless typhoon. I am ready to celebrate, to pat myself on the back and gloat a bit when I spot Adam's bag on the luggage rack above us. 'Is that yours?' I ask him in a voice tinged with disbelief. 'Yeah, why?' he replies. Adam's case is a replica of the Gladstone-style bag, the gripsack that Nellie Bly carried around the world. I need a minute to take it all in before attempting to explain this uncanny coincidence. Adam seems a little baffled by my excitement. I power up my tablet so he can see for himself. Among the many images I keep in my Nellie Bly photo file is one of her gripsack. Adam glances at the screen, then takes a hold of the tablet for a closer examination. It is true; his bag is the same as Nellie's. Until that moment, he had never heard of Nellie Bly. The swanky travel bag he chose simply to express his sense of fashion now has a historic connection.

My two-hour train journey with Adam on the slick and speedy 'iron horse', Hong Kong's Intercity Express, is the less enticing alternative to an overnight river trip beneath the starry skies. I planned to travel by boat, but Kalmaegi robbed me of the same Pearl River pleasures Nellie experienced. Indeed, I am lucky to be going to Canton at all.

1. Nellie Bly in her legendary travel attire. (*Courtesy of the Library of Congress, LC–USZ62-59924*)

2. 'Nellie Bly Bids Fogg Good Bye' trading card. (*Courtesy of the Alice Marshall Women's History Collection, Ephemera and Artifacts, Accession No. AKM 91/1.1. Archives and Special Collections at the Penn State Harrisburg Library, Pennsylvania State University Libraries*)

3. Author's daughter Acadia on the steps of the Maison Jules Verne in Amiens, France. (*Photo by author*)

4. Grand Oriental Hotel, Colombo, Ceylon, circa 1890s. (*Courtesy of Lankapura, www.lankapura.com*)

5. Author at Full Moon Poya celebrations in Colombo, Ceylon.

6. The replica statue of Sir Stamford Raffles, Singapore. (*Photo by author*)

7. Author touring temple in Singapore's Little India.

8. Hong Kong Harbour, 1889. (*Digital image courtesy of the Getty's Open Content Program*)

9. Happy Valley Cemetery swamped by skyscrapers, Hong Kong. (*Photo by author*)

10. Firing of the Noonday Gun, Hong Kong. (*Photo by author*)

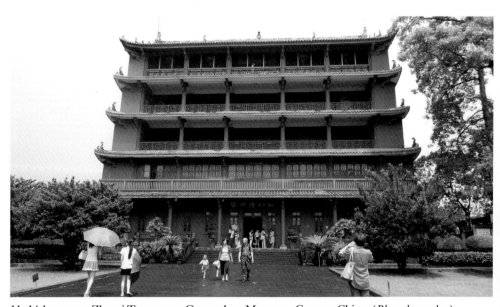

11. 14th century Zhenai Tower, now Guangzhou Museum, Canton, China. (*Photo by author*)

12. Temple of the 500 Gods, Canton, China, in Nellie's time. (*Courtesy of the Library of Congress, LC-DIG-ggbain-09935*)

13. Today's Temple of the 500 Gods, Canton, China, now called Hualin Temple. (*Photo by author*)

TEA SERVICE AT THE GRAND HOTEL, YOKOHAMA, JAPAN.

14. Tea at the Grand Hotel, Yokohama, Japan. (*Postcard from the collection of the Kanagawa Prefectural Museum, Yokohama, Japan*)

15. Sangedatsumon Gate, Zojoji Temple, Tokyo, Japan. (*Photo by author*)

16. Lap of the Great Buddha, Kamakura, Japan. (*Photo by author*)

PRESENTING THE GLOBE-GIRDLER A GOLDEN GLOBE.

THE ARRIVAL IN PHILADELPHIA.

AROUND THE WORLD IN SEVENTY-TWO DAYS AND SIX HOURS—RECEPTION OF NELLIE BLY AT JERSEY .CITY ON THE COMPLETION OF HER JOURNEY—From Sketches by C. Bunnell.—[See Page 7.]

17. Nellie Bly's triumphant arrival and reception in Jersey City after circling the world in seventy-two days. (*Courtesy of the Library of Congress LC–USZ61-2126*)

18. 'She's Broken Every Record!', front page of *The New York World*, 26 January 1890.

19. Nellie Bly's Round
the World Game.
(*Courtesy of the Library
of Congress LC-DIG-
ppmsca-02918*)

20. Staircase at the New York City insane asylum where Nellie Bly went undercover. (*Courtesy of the Library of Congress, HABS NY,31-WELFI,6-*)

NEW-YORK CITY ASYLUM FOR THE INSANE (WOMEN), BLACKWELL'S ISLAND.

21. The New York City insane asylum in Nellie Bly's day. (*British Library, accession number: HMNTS 10413.g.22*)

DEDICATED JUNE 22, 1978
TO
NELLIE BLY
ELIZABETH COCHRANE SEAMAN
BY THE NEW YORK PRESS CLUB
IN HONOR OF
A FAMOUS NEWS REPORTER
MAY 5, 1864 – JAN. 27, 1922

22. Author laying roses at Nellie Bly's headstone, Woodlawn Cemetery, New York City. (*Photo by Alice Robbins-Fox*)

23. Dolly Lackey McCoy, author, and Arnold Blystone at millstone memorial to Nellie Bly in Cochran's Mills, Pennsylvania. (*Photo by David Stanton*)

24. Nellie Bly's childhood home in Apollo, Pennsylvania with historical marker in front. (*Photo by author*)

25. Concept Rendering by Amanda Matthews for *The Girl Puzzle* Installation on Roosevelt Island, NYC © 2020. (*Artist, Amanda Matthews, www.prometheusart.com*)

Recent happenstances have me pondering. How did a taxi driver inexplicably emerge in the midst of a raging typhoon? Why, on an empty train, am I seated next to someone carrying a replica of Nellie Bly's gripsack? These thoughts mystify me. How many people are on this journey? Is Nellie by my side? Or perhaps some of her invincible spirit has rubbed off on me.

Chapter 9

In Which Nellie Spends Christmas in Canton

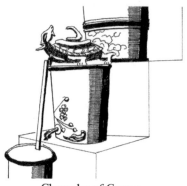

Clepsydra of Canton

Canton (Guangzhou)
24–25 December 1889

'I felt I was a long ways off from my own dear land; it was Christmas day, and I had seen many different flags since last I gazed upon our own.'

Nellie Bly

T he SS *Powan* docked in Canton before daybreak on Christmas day. Nellie and her fellow passengers enjoyed breakfast on board the river steamer before transferring to sedan chairs for a tour of the port city led by Ah Cum, Canton's pre-eminent guide. So acclaimed was Ah Cum that some considered him one of the local 'sights'. The English-speaking tour guide had a reputation for knowing precisely what visitors should see and do in his city. Indeed, Ah Cum had once worked in the household of the Venerable John Henry Gray, inaugural Archdeacon of Hong Kong and author of the 775-page *Walks in the City of Canton*. Ah Cum held a monopoly on British visitors – royalty, statesmen and other notables. His services could even be engaged through the local office of Thomas Cook.

'A Merry Christmas,' Ah Cum greeted his group as he boarded the SS *Powan* that morning to supervise the picnic lunch they would carry with them. Nellie admired his black-beaded shoes and their stark white soles, his quilted silk 'smoking jacket' and navy leggings, and his coal-black pigtail reaching down to his heels that he tied with a black silk tassel. The guide's inch-long fingernails were testimony to his status. Ah Cum's character, cunning and connections were the tools for his job, not his hands. Aside from the $1 charge per tourist (£21/$26 today), the shrewd guide pocketed a percentage of the purchases made by his groups to the only shops he encouraged them to enter. They also paid a fee for the sedan chairs and bearers he had arranged to carry them. Nellie could not help noticing that Ah Cum's sedan chair was far superior to those she and the others were mounting. His chair was jet black and adorned with fringes and tassels; its polished poles were capped with brass knobs. The bearers were equally resplendent – dressed in white linen decorated with red bands – looking like 'circus clowns,' she wrote. Nellie's group would travel in plain willow chairs with ordinary covers that often obscured their views. Barefoot, perspiring Chinamen clad in shabby navy blue shirts and trousers stood in position, two holding the poles in the front, and one at the back.

Leaving the dock, the group embarked on Ah Cum's 'deluxe' tour that would lead them through markets and temples into the more chilling side of this unforgiving city where torture and executions were commonplace. As they set off, the streets were so cramped and so narrow that Nellie thought they were being carried through an outdoor market. 'It is impossible to see the sky, owing to the signs and other decorations, and the compactness of the buildings,' she wrote. 'When Ah Cum told me that I was not in a market-house, but in the streets of the city of Canton, my astonishment knew no limit.' Once clear of the port, the astute guide started with the familiar, weaving his group through Canton's teeming and malodorous streets, into the exclusive international enclave of Shameen Island (today's Shamian Island). He wanted them to experience a classic European environment before he introduced them to his city's more alarming side. Shameen Island was, and remains today, an oasis of tranquillity and charm in a frantic city. Nellie wrote of Shameen as 'green and picturesque, with handsome houses of Oriental design, and grand shade trees, and wide, velvety green roads … broken only by a single path, made by the bare feet of the chair-carriers.' Except for the bearers' footpaths, the same could be said of the island today. Parked like a ship alongside a frenetic mainland, Shameen was a concession from the Chinese to the Europeans between 1859 and 1946. When Nellie was here it was controlled by the British (sixty percent) and the French (forty percent). Two ornamental white stone bridges still link Shameen to the mainland – the West Bridge or Bridge of England, and the East Bridge, the Bridge of France. I cross the Bridge of England to join the small island, a treasure chest of colonial architecture – foreign consulates, churches, banks, hotels and parks – all linked by shaded promenades. Foreign diplomats and merchants recreated a Western lifestyle here, leaving their stamp on this tiny slice of Europe in the Far East. More than 150 colonial buildings were constructed in the European styles of the day – gothic, neo-baroque, and neoclassical – all garnished with oriental flourishes. Many still stand.

On Shameen, multi-rooted banyan trees front the landscaped gardens of mansions hugged by wrap-around verandas. Grecian columns, classical porticos and pediments, sash windows and friezes abound on this former sandbank that can be traversed in fifteen minutes from top to bottom. Leafy branches meet and form 'bridges' over paved walkways bordered by flower beds and fountains. Wearing its colonial legacy with pride, even flaunting it, the little island now signposts these fine buildings, noting their roles before occupants flocked to Canton's ultra-modern Central Business District. Historical markers with maps and descriptions invite visitors to explore the area now celebrated as a heritage site by the Chinese government. Indeed, the markers state that the island 'belongs not only to Guangzhou, but also to China and the world as a whole'. It is another world in another time; late nineteenth century pocket-sized Europe in the People's Republic of China. The deep-earth aromas of the local cuisine and piercing exchanges in tonal Cantonese are occasional reminders that I am in China, along with the ubiquitous Flying Pigeon bicycles and preponderance of cats. Predictably, both Nellie and I feel an affinity with Shameen. This wing-shaped island is my headquarters during a three-day stay in Canton at the Customs Hotel, a neoclassical building set amid a grove of trees on Shameen's main street. This European 'flavour' is like comfort food for me after ten days in Asia. I am sorry that today's global retail chains such as Starbucks and 7-Eleven convenience stores have invaded. But it is reassuring that the architectural gems of the past, unlike many other historic Chinese sites, survived China's Cultural Revolution. On Shameen, I can pretend to live in Nellie's times.

The United States built its consulate here in 1873 with the concurrence of the British government. Sixteen years later, Nellie arrived on the scene. This is where she caught her first glimpse of 'home' since her voyage began almost six weeks earlier. As Ah Cum's entourage proceeded along the island, Number 56 Main Street came into view. Nellie's heart leaped. There it was – the American flag – thirty-eight stars on a midnight blue background, bounded by thirteen red and white stripes, waving before her. 'Here for the first time since leaving New York, I saw the stars and stripes,' she wrote. 'I felt I was a long ways off from my own dear land; it was Christmas day, and I had seen many different flags since last I gazed upon our own.' The further Nellie travelled from home, the more she appreciated it. 'The moment I saw it floating there in the soft, lazy breeze I took off my cap and said: "That is the most beautiful flag in the world, and I am ready to whip anyone who says it isn't."' Silence prevailed. 'No one said a word. Everyone was afraid,' she wrote. 'I saw an Englishman in the party glance towards the Union Jack, which was floating over the English Consulate, but in a hesitating manner, as if he feared to let me see.' Nellie and her party were welcomed to the American consulate by Charles Seymour, US Consul in Canton from 1884 to 1897. They stayed just long enough to pay their respects, but he made quite an impression on Nellie. 'Mr. Seymour is a most pleasant, agreeable man, and a general favorite. It is to be hoped that he will long have a residence in Shameen, where he reflects credit upon the American Consulate.' Seymour remained in post a further eight years. Departing from the consulate, Ah Cum's procession passed the Anglican Church, the athletic club with its tennis courts, the boat house, all of which still exist, before crossing back into the maelstrom of Canton. As the parade of sedan chairs continued into the centre, their occupants encountered a potpourri of smells – rotten fish, foul water and penetrating spices – along with the incessant chatter of shopkeepers jammed along noodle-thin lanes. After the serenity of Shameen, the chaos of Canton was quite a jolt. 'What a different picture Canton presents to Shameen,' Nellie wrote.

Winding through the markets, Nellie noticed altars in the back of shops, 'gay in colour and expensive in ornament' and wooden desks at the entrance to stalls where the bookkeepers wore huge tortoise-shell rimmed glasses which 'lend them a look of tremendous wisdom.' She was intrigued by the local people – their clothing, their pigtails and chignons, their customs. Nellie also realised that she and her travelling companions presented quite a spectacle to those not accustomed to seeing foreigners, especially someone like her. In 1889, an American woman abroad was quite a rarity. Customers stepped out of shops to stare and sometimes touch her. Nellie's gloves attracted the most attention; clothing for the hands was unheard of in China. 'They gazed upon them with looks of wonder,' she wrote.

Then and now, markets in Canton tend to be clustered according to their wares: medicines and herbs, fruit and vegetables, flowers and birds, even coffins and shrines. Notorious Qingping market, along the Pearl River across from Shameen, is an exception. Vendors seem to sell everything here. With Nellie in mind I wander through, trying not to look too closely in fear of seeing something I do not wish to recognise. An acrid stench of manure, mould and formaldehyde pervades. Live snakes, kittens, rabbits, monkeys and other creatures huddle and squirm inside stacked metal cages. In neighbouring stalls their dead equivalents are skinned and preserved in brine. Grubby plastic tubs teeming with live baby scorpions sit alongside polythene bags stuffed with deer antlers, dried seahorses, tiger paws and chicken feet. Large musty jars balanced on makeshift shelves are packed with twisted, bone-like contortions, hairy roots, knuckle-shaped balls, shrivelled skins and other suspicious items. I am not sure if they are food, pets, remedies, drugs or even legal. I hope they are not human.

After the markets, Nellie was anxious to see Canton's renowned execution ground. A grisly, but popular tourist attraction in its day, the ground was chronicled in travel journals by inquisitive westerners probably guided there by Ah Cum. Arriving just outside the city, Nellie's 'little train' dismounted their sedan chairs and followed Ah Cum down what looked like a 'a crooked back alley in a country town'. It was only seventy-five feet long and twenty-five feet wide, narrowing down at one end like an isosceles triangle. This was the execution ground, Ah Cum announced. Glancing around what seemed like a rather nondescript site, Nellie noticed a 'very red' patch of ground. 'Human blood,' Ah Cum proclaimed, as he kicked the scarlet-soaked earth with his white-soled shoe. The sticky red evidence of a mass beheading was festering in the winter sun before them, emitting a sickly metallic odour. Twenty-four hours earlier, eleven men, their hands crossed and bound, knelt on the now-stained soil and hung their heads ready for the executioner's sabre. One by one, their lives were cut short in a filthy back alley where justice had no place. One foreigner who had witnessed an execution there just a month earlier wrote that after the terror and torture they had endured, the prisoners actually welcomed their ghastly deaths. Executioners, grasping a yard-long sabre with both hands, dealt blow after blow after blow. Thirty-three heads could be severed in as many minutes. Ah Cum said that ten to twenty criminals were usually executed at a time, estimating an average of 400 beheadings a year. During the Taiping Revolution in 1855, 55,000 rebels were decapitated in this grim alley. Rivers of blood formed crimson ponds and the air reeked of decomposing heads, Ah Cum told his group. While he was speaking, Nellie noticed some roughly hewn white pine crosses propped against a high stone wall. She assumed they had a religious significance. Now the horror levels rose higher. These simple crosses were employed for the most terrifying punishment in history. They were the instruments of death by the long, lingering, excruciating torture known as *lingchi*. The guide explained that the condemned are bound to the crosses and cut to bits so deftly that they are entirely dismembered and disembowelled while still alive. *Lingchi*, which translates loosely to 'slow slicing', 'lingering death' or 'death by a thousand cuts', was not officially outlawed in China until 1905, twelve centuries after it originated.

'A shiver waggled down my spinal cord,' wrote Nellie after listening to the guide's grisly descriptions. But they didn't seem to stem her gruesome curiosity. 'Would you like to see some heads?' Ah Cum enquired as if offering a look at bronze busts in a museum. Nellie did not take him seriously. As a New Yorker, she knew that guides often exaggerated. Calling his bluff, with steel in her voice she said, 'Certainly, bring on your heads!' At Ah Cum's instruction she tipped a man who proceeded to one of the lime-filled clay barrels pitched against the wall alongside the wooden crosses. He reached his hand deep inside, rummaged around, and pulled out a severed human head! At least ten others were smothered in lime below it, Ah Cum said. Nellie may have been repulsed and horrified, but she did not show it. Her explanation: 'Chinamen are indifferent about death; it seems to have no terror for them.' We might say the same about her.

The trail of terror continued as the foreigners were carried to the nearby jail and court. Nellie was surprised to find the jail unlocked. There was no need for security; the half-dead prisoners could not escape. Their bleeding necks were choked by hefty wooden boards that prevented them from lying down or resting their heads, leading eventually to death by exhaustion. A real-life torture chamber, the nearby courthouse held heinous tools for punishment: thumb screws, pulleys to hang prisoners by their thumbs, split bamboo for whipping. Ah Cum regaled them with tales of the agony that simple bamboo can inflict. It is at its most excruciating when used as a human skewer. Standing with legs astride, the condemned is staked to the earth above a bamboo sprout that reminded Nellie of 'delicious asparagus'. But this sage-green shoot is as tough as iron and grows rapidly – impaling its victim, in intolerable pain, inch by inch.

The inhumanity they had already witnessed was not enough for Nellie. She had a 'great curiosity' to travel to the leper village placed northeast of the city walls, well out of the city centre. Even before their sedan chairs entered the colony, the foreigners were engulfed by the putrid smell of rotting flesh. They masked the stench by smoking. 'Ah Cum told us to smoke cigarettes while in the village so that the frightful odors would be less perceptible,' she wrote. While their noses were partially spared, the foreigner's eyes could not help but witness the unbearable wretchedness. 'The lepers were simply ghastly in their misery,' wrote Nellie. 'It is useless to attempt a description of the loathsome appearance of the lepers. Many were featureless, some were blind, some had lost fingers, others a foot, some a leg, but all were equally dirty, disgusting and miserable.'

Nellie's morbid curiosity disturbs me. Even as a fellow journalist, I am uneasy with her cold-blooded captivation with the macabre. Perhaps her fascination arose from her own experience at the brutal hands of the nurses and doctors at the lunatic asylum on Blackwell's Island when she went undercover for *The New York World*. Nellie's harrowing exposés made her name as a reporter and set the stage for the historic global voyage that led her to Canton, and into the leper colony that I am trying to locate 125 years later.

Even after scrutinising nineteenth-century maps, I cannot find any traces today of the leper village that consigned vulnerable human beings to a living hell. Equally absent is physical evidence of the execution ground, jail and court. To be honest, I am relieved. Canton's bloody execution ground, where tens of thousands of heads were severed, has gone underground. Executions in China are no longer public and are generally carried out by lethal injection or a bullet to the back of the head. China executes more people every year than the rest of the world combined. Estimates of the death toll run into the thousands, according to human rights organisation Amnesty International, but the number put to death each year in China is a state secret. And so, it seems, are the locations of the barbaric places that Nellie visited with Ah Cum on Christmas day 1889.

She saw them all, at her own alarming request. I am here to follow her. It is easy to shadow her in the former colonial enclave of Shameen. But once off the island, the challenge is immense. If Shameen is a treasure hunt, Canton is a scavenger hunt. Few clues lead even to remnants of the prominent sites that beat at the heart of Canton when she was here. Many perished when Mao Zedong unleashed the Cultural Revolution, in the appalling decade between 1966 and 1976 when China was purged of much of its heritage and as many as two million of its people.

Nellie experienced a different China, and I am struggling to find it. I am not alone in my search. The obliging young receptionists at the Customs Hotel where I am staying are quick to whip out their phones and scour Chinese cyberspace on my behalf. They also translate the names of the sites I am seeking into Chinese characters so I can show them to local people when asking directions. This comes in particularly handy when tracking the religious temples visited by Nellie. With 800 shrines in the city in that era, Ah Cum sifted out two of the best for his group – the Temple of Horrors and the Temple of the 500 Gods. Today they have different names making it almost impossible to locate them. A twenty-first century equivalent of Ah Cum would come in handy, but the hotel receptionists and the World Wide Web will have to suffice. They do. Online, we unlock the name and location of the Temple of Horrors through a letter written by an American sailor, Jay Floyd Cole, who described the ghastly temple to his family in a letter during a visit to Canton in 1883 when he was twenty-two. In 2010 Cole's grandson donated his grandfather's correspondence to the Museum of Guangzhou, and the mystery surrounding the Temple of Horrors was solved. It is today's Cheng Huang temple, also known as the City God Temple, constructed in 1370. Like many found in China's cities and provinces, the temple was erected during the early Ming dynasty to worship the gods who protect the local area, but also to terrorise its residents. In an approach that remains all too familiar in today's China, fourteenth-century Emperor Zhu Yuangzhou said: 'I build Cheng Huang temples to make the public fear. They will behave themselves if they fear.' Inside Canton's Cheng Huang temple, hideous god statues unleashed their fury in lifelike spectacles of butchery and bloodshed. Sinners were whipped, beheaded, boiled in oil and sliced in half lengthwise in graphic portrayals that more than justified the Temple of Horrors' epithet. Nothing was left to the imagination in the Ming emperor's determination to terrorise his citizens into submission. Left to rot for centuries, the Temple of Horrors was salvaged in 1993 and listed as a historic site. Cheng Huang temple was given a new life, a new look and a new purpose as a showcase for Chinese history and culture just in time for the 2010 Asian Games hosted here. Eighteen fierce god effigies inflicting fear and torture were demolished and replaced by three benevolent deities charged with guiding the public towards goodness. Neither Jay Floyd Cole, nor Nellie Bly or even Ah Cum would recognise the transformed Temple of Horrors today. It stands in the shadows of surrounding high-rises in the commercial centre of Canton. I approach it through a monumental white stone gate with a trio of portals crowned by winged roofs that look like dragon boats. Inside the main hall, three larger-than-life gods are seated on thrones, their gentle smiles emerging from Fu Manchu moustaches. A golden lacquer mural almost the size of a tennis court extends across one wall. Emblazoned across the glowing expanse, 132 gods from Chinese mythology float in a gilded heaven inhabited by peacocks, storks and dragons. A soaring mahogany-ribbed ceiling is supported by enormous black pillars embellished with gold Chinese characters. I am joined inside by local worshippers and tourists who have come to greet the gods, pray for favours and admire the temple's sacred art and culture. Here at Cheng Huang temple, an incense-infused reverence, not fear, prevails among those of us in attendance.

Not so in Nellie's day. As she approached the Temple of Horrors, mournful wails and desperate cries for alms emanated from the twisted, diseased bodies spilling across its dirt-strewn steps. The human misery on the steps outside led to an 'exhibition of human monstrosities inside,' according to Nellie. 'A filthy stone court was crowded by a mass of humanity – lepers, peddlers, monstrosities, fortune tellers, gamblers, quacks, dentists with strings of horrid teeth, and even pastry cooks.'

Repelled by the blatant filth and misery at the Temple of Horrors, Nellie much preferred the second temple she toured during her 'flying visit' to Canton. The Temple of 500 Gods, tucked deep inside the vibrant jade market on the city's west side, was her favourite. With 500 golden life-size statues as clues, locating this temple should be a cinch. But it is not. I have two monochrome photos of the gods inside the nineteenth-century temple, but these snaps hinder rather than help the search. The shrine in my photos no longer exists; it has been replaced. It is called Hualin Temple and I can get there on foot in thirty minutes. Nellie's golden 'gods' are actually *arhats* or Buddhist saints. The temple and every one of the 500 original saints placed there in 1851 were demolished in Mao's Cultural Revolution. Thirty years later it was reconstructed, 500 brand new saints were installed, and today's Hualin Temple was opened to the public once again. A metal plaque outside the temple provides this brief, but enlightening history, etched in Chinese and English. Just like their predecessors, each of the 500 gleaming saints strikes a unique pose and expression. Many clasp canes, cats, beads, babies, vases; all are seated, some cross-legged. These saints, all male, are deified warriors, heroes, sages and apostles of the Buddhist faith. Marking his visit to China in the thirteenth century, Marco Polo is among them. I find him surrounded by rose and emerald lotus flower lamps. He wears a pork-pie hat and a flowing cape; wooden shoes peek out under billowing harem pants. Around him, row upon row of saints rest inside glass cases framed in dark wood decorated with gold filigree. Above their 'framed' peers, hundreds more *arhats* pose on shelves reaching up to ceilings painted with swirling mandalas. 'It looks more like a gallery of sculpture than a place of worship with these long lines of solemn looking figures staring each other out of countenance century after century!' wrote photographer John L. Stoddard of the temple in *Glimpses of the World; a Portfolio of Photographs of the Marvelous Works of God and Man*, which was published in 1892 . 'They certainly are not praiseworthy as works of art, yet incense is burnt constantly before some of them, and the air is heavy with that pungent perfume.' This was the temple that Nellie saw. Except for the ever-present 'pungent perfume', the twentieth-century temple bears little resemblance to its ancestor. Today's golden saints are much jollier.

Tracking down the Hualin and Cheng Huang temples that Nellie visited has required research, time and determination which must be equally applied to sites I may never find. I decide to focus on my two remaining priorities – an ancient bronze water clock described in great detail in Nellie's short book, and the Temple of the Dead where she ate her lunch on Christmas day.

Nellie's own account of the Temple of the Dead provides no clues as to where I might find it. She writes only of being behind a high wall and coming upon a 'pretty scene' of a pond overhung with tree branches that 'kissed the still water where stood long-legged storks' like those painted on Chinese fans. From there, according to Nellie, Ah Cum led them into a room with chairs and tables shut off from the court by a large carved gate. It is here the foreigners were served a picnic lunch on 25 December 1889. It was a normal day in China. With the exception of Shameen and the US Consulate, Nellie had spent Christmas morning on a chilling excursion through Canton's grisliest sites. 'As we left the leper city I was conscious of an inward feeling of emptiness,' she wrote. 'It was Christmas day, and I thought with regret of dinner at home.' By the time Nellie sat down for her yuletide luncheon, it was about midnight in New York. As Ah Cum's group began their picnic, a chorus of chanting 'to the weird, plaintive sound of a tom-tom and shrill pipe' drifted into the room. Once Nellie had 'less appetite and more curiosity', she asked the guide for an explanation. They were inside the Temple of the Dead, Ah Cum replied. Outside, a death march was taking place. The sounds they could hear were part of the Chinese funeral rites of passage. 'But that did not interfere with the luncheon,' wrote Nellie. She was 8,000 miles from her home in New York City, celebrating the western world's most festive holiday amidst mourners and decomposing bodies in the Temple of the Dead. I really want to find this temple. Pre-trip research revealed no references to the Temple of the Dead or anything resembling it. Even here in Canton, where it may still exist, I come up empty-handed. Reluctantly the Customs Hotel receptionists and I abandon Nellie's Christmas luncheon venue; but we are not yet ready to give up our search for the water clock that so fascinated her. I have identified it as the Clepsydra of Canton, one of the world's most extraordinary clocks. It is bafflingly elusive. Designed in AD 1324, this clock, which designated the time by water drops over a twelve-hour period, once stood against the city's northern gate. That is where Nellie saw it – 'four copper jars about the size of wooden pails, placed on steps, one above the other. Each one has a spout from which comes a steady drop-drop,' Nellie wrote. 'A float on the lowest bucket indicated the approximate time of day.' The ancient northern city gate no longer exists and neither, it appears, does the venerable clock. Like the Temple of the Dead, it seems to have been airbrushed from history. After several false and frustrating leads, I feel forced to relinquish the quest.

Believing I have found all the other existing sites in Canton, I decide I can take a break from the relentless Nellie Bly trail, leave yesterday's Canton behind, and be a tourist in modern day Guangzhou for a morning. I head downtown to explore China's largest urban park, Yuexui, a jade carpet flowing over undulating hills in the midst of a roaring city. Light autumn showers cast a sheen on the lush foliage as visitors unfurl pretty umbrellas – floral, striped and spotted – that sprinkle the park like flowers. Raindrops release freshness into air usually overwhelmed by pollution. I wander through the gardens happy to be outdoors, even in the rain. As the cloudbursts intensify, I veer upwards towards the park's renowned centrepiece, the fourteenth-century green-tiled Zhenai Tower, now home to the Guangzhou Museum. Perched atop Yuexui Hill, Zhenai Tower was known as the Five-Story Pagoda in Nellie's day. I duck inside to avoid a drenching and uncover a museum charting the port city's history over 5,000 years. Each of the five storeys represents a millennium; one thousand years per floor. I start at the top, in the nineteenth century, to gain context for Nellie's era. Following the staircases down through time, I cruise through the past, skimming the centuries like the sections of a Sunday newspaper.

Suddenly I put on the brakes. I anchor my feet to the ground. There it is. Nellie's water clock! Right in front of me in all of its ancient glory. Stationed on a stair-like brick platform against a life-size photograph of the clock *in situ* beside the northern city gate, the elusive clepsydra endures! Four bronze buckets step down in succession; the second bucket is crowned by an alluring, mythical bronze creature – part goat, part turtle and part snake. Stunned and elated, I have the inexplicable, but very real feeling, that I have been led to it. Nellie again? Less than twenty-four hours since I relegated it to obscurity, this mysterious clock presents itself. Rescued from the northern city gate, it has taken up residence in the city's history museum that I would never even have entered if it had not been raining. Both Nellie and Mother Nature are on my side. My head is spinning. Out comes the camera as I try to capture this revered clock.

I take photo after photo, trying to avoid the stop-sign red universal 'do not touch' signs obscuring this sacred timepiece. A guard approaches. Using my best body language, mostly hands and eyebrows, I attempt to request permission to remove the signs. Just for a few seconds. Not a chance, comes the silent reply, written in pursed lips and rigid arms crossed firmly over her chest. She returns to her station, just twenty yards away. I watch her settle into it. Praying that she is not looking – and doing just what Nellie would do – I gather up the five red signs and quietly pile them into a nearby corner. The guard is back at my side immediately. I am forced to return the signs to their rightful, but infuriating, positions. I mime my disappointment. Contorting my body into a range of poses, I try to capture the clock despite the signs. But it is impossible. The guard re-appears. She shifts her eyes to the left. It is a signal. I think she is letting me know that she will remove the signs for me, but I had better be quick. The deal is done. I snap away, unhindered by the tiresome signs. I am grateful and I think she feels pleased too; but not enough to let me take her picture in front of the water clock. I have found and photographed the Clepsydra of Canton, the rain has eased. I take my leave wondering if Nellie is still with me.

Ah Cum did not lead his group to my next stop, the Museum of the Mausoleum of the Nanyue King. The 2,000-year-old tomb of King Zhao Mo was not discovered until 1983, when excavations were underway to build a commercial development on the hilltop where the museum now stands. In one of China's most momentous discoveries, bulldozers exposed seven burial chambers brimming with more than 10,000 funerary artefacts from the Han King's tomb. Among the ancient treasures, archaeologists uncovered a chariot, vessels of gold, silver and bronze, and most remarkable of all, the king's elaborate burial suit made of 2,291 tiles of precious jade pieced together with maroon silk. They also unearthed evidence that the king was not alone in his tomb. Fifteen living people joined him in this underground 'palace' to serve him in his afterlife. The king died of illness in 122 BC, unlike his loyal attendants – a guard, a musician, concubines, cooks and servants – who were put to death. Descending stone stairs into the burial chambers, I follow a narrow passage flanked by slab-like walls. It smells like a cave. A map mounted nearby shows where the sacrificial bodies were positioned in the king's tomb. I am inspecting it when suddenly I hear my name. Who could possibly know me here in Canton, let alone in the burial place of King Zhao Mo? It is Adam, my seatmate on the train from Hong Kong to Canton, the one with the lookalike Nellie Bly bag. He is running late so we only have time to hug, take a selfie together, and then go our separate ways, happy that our paths have crossed once again.

Walking distance from the macabre mausoleum, the Canton Orchid Garden beckons. I have seen more orchids blooming in the floral section of a UK supermarket. It is not the season; the majority of the garden's 200 species are blossomless. Even so, the setting offers a sea of green on a rainy day that is being transformed into a sauna as the sun arrives. Winding paths and stepping stones lead me through bamboo stands and small groves of saplings. Stone bridges cross willow-fringed ponds towards landscaped gardens and greenhouses. I watch an elderly gardener crouched on a lawn like a backwards 'Z'. She is sitting on her own heels removing weeds one by one with a hook, comfortable in a stance that would send Westerners straight to an osteopath. After watching her for a while, I follow a trail to a secluded wooden teahouse that seems to float on the lakeside. Sweeping pine and cypress boughs forge a camouflage that deepens the stillness here. Tea is served on the deck outside. I find a seat right at the edge for a close-up view of the golden carp and speckled koi swirling in the water below.

The afternoon well behind me, I decide to end my visit to Canton in the same way that Nellie began hers – on a ship. Her overnight journey on the SS *Powan* along the Pearl lasted eight hours; my night cruise will ply the river for seventy minutes. With Nellie and the *Powan* in mind, I board the *South Sea God*, a massive dragon boat-cum-galleon ablaze in red and yellow neon. An illuminated dragon's head with a long and billowing mane clings to the prow where a carved mermaid or unicorn might be found on a European ship. This Chinese figurehead, symbol of power, strength and good luck, will lead us along the neck-stretching cityscape of geometrical superstructures rising from Guangzhou's downtown. The Pearl River is the silky black ribbon that connects them all. Looping lasers paint the skyline in dayglo colours. Flaming rainbows unfurl across bridges as the *South Sea God* glides through the night. The highlight of this pageant of lights is the Canton Tower, bathed in psychedelic hues as it gently twists 1,982 feet towards the moon. The Asian illuminations draw me like a moth; it is as if I am at the centre of an enormous kaleidoscope. On her Pearl River journey aboard the SS *Powan* to Canton, Nellie embraced the stars. Tonight I bask in the lasers.

Chapter 10

In Which Nellie Falls in Love with Japan

Lion's head water hydrant, Yokohama

Yokohama
2–7 January 1890

'In short I found nothing but what delighted the finer senses while in Japan.'

Nellie Bly

Nellie wrote about Japan like a lovesick adolescent. She was infatuated. Her somewhat cynical nature melted in the Land of the Rising Sun. 'If I loved and married, I would say to my mate: "Come I know where Eden is and ... desert the land of my birth for Japan."' This from a fiercely patriotic Nellie who defended her beloved USA at every step along her journey. Accolades for the Japanese dappled her text like sunlight – charming, sweet, happy, cheerful, graceful, pretty, artistic, obliging and progressive. She adored everything about the country, especially the people – their cleanliness, their grace, their clothing, their children. 'In short I found nothing but what delighted the finer senses while in Japan,' gushed an otherwise snide Nellie Bly.

She fell in love with a country that had closed its doors on the outside world for more than 220 years until 1853 when US Naval Commodore Matthew Perry stormed in with warships to force them open. This was gunboat diplomacy. When, as agreed, Perry returned a year later to end Japan's self-imposed isolation, he strode ashore where his warships were anchored. This was the moment when a humble fishing village hugging the Pacific coast earned a place in world history. The settlement of Yokohama was the site for the signing of the Treaty of Peace and Amity on 31 March 1854. The treaty, also known as the Convention of Kanagawa, would transform Japan from a closed, feudalistic society to an open, modern nation state.

Thirty-six years later, when the SS *Oceanic* sailed in with Nellie on board, Yokohama had just become a municipality. Now it is Japan's second largest city and one of its leading ports. Yokohama was then, and is now, a crossroads where east meets west. If Japan was a house, this coastal city would be its front door. It is home to the nation's first railway, linking the port to the capital city of Tokyo, twenty miles away. In Nellie's day the railway journey took an hour; today it is a twenty-five minute commute for me on the Tokyu Toyoko Express.

*   *   *

When Nellie stepped ashore in Yokohama, she began a five-day love affair with Japan. 'Rain the previous night had left the streets muddy and the air cool and crisp,' she wrote. But the sun was 'creeping in' as she arrived in this brand new city she described as 'having a cleaned up Sunday appearance'. I arrive on a Saturday. An autumn drizzle turns the sidewalks into silver as I exit the station to begin my search for Nellie. It soon becomes clear that I have entered a Yokohama that she would not recognise. Her colonial city is gone, her footsteps have vanished. Almost all vestiges of the late nineteenth and early twentieth centuries have disappeared. They were swallowed by the Great Kanto Earthquake of 1923, which robbed the port of its livelihood, its heritage and much of its population in the deadliest natural disaster in Japanese history. The devastating story unfolds for me inside the tourist information centre near the station.

At noon on 1 September 1923, a 7.9 magnitude earthquake rocks an unsuspecting Tokyo-Yokohama area preparing for lunch. Within minutes a seven-storey tsunami hurls out of the sea, unleashing a chain of monstrous waves that sweep thousands to their death. The quake ignites wooden houses like matches; they tumble like dominoes across the two cities. Melting tarmac glues victims to their tracks as the death rate soars in a sea of flames. The iconic Grand Hotel, where Nellie stayed, collapses, crushing hundreds. American journalist Henry W. Kinney is watching the disaster unfurl. He compares Yokohama to a 'gigantic Christmas pudding, over which the spirits were blazing, devouring nothing. For the city was gone.' One British expat ponders, 'Will Yokohama be a great city again? I think not.' In Kamakura, thirty-seven miles from the quake's epicentre, the colossal 121-ton Great Buddha statue, visited by Nellie, lurches two feet from its 600-year-old foundation.

The Great Kanto Earthquake levelled the nation's two largest cities, killing more than 140,000 people and leaving 3.25 million homeless. The disaster plunged Japan into a deep trauma … and still does. The quake's 1 September anniversary is Disaster Prevention Day across the country, where citizens rehearse evacuations, firefighting and rescue operations – ready to tackle the next earthquake, tsunami or typhoon.

I am not prepared for the post-quake city. I am not sure where to start. All I can do is dodge the drizzle and try to unravel the past, seeking out the footprints of sites lost more than eighty years ago. I begin at the waterfront, a short walk from the tourist office. Back then the Grand Hotel, said to be *the* place to stay, commanded the eastern end of the port. It boasted of being the 'the most-liked hotel in the world' where the 'superlative excellence of American ideals' merged with the 'Japanesy charm of the Opulent East'. Nellie admired the airy rooms, the exquisite views, the splendid food and the excellent service. What she despised were the rodents. 'Barring an enormous and monotonous collection of rats,' she wrote, 'the Grand would be considered a good hotel even in America.'

A tourist destination of a different sort now stands on the Grand Hotel's waterfront site. Ten thousand dolls representing 140 countries reside where the fashionable hotel once hosted the 'surging tides of travel ceaselessly sweeping round the world to and from the Orient and the Occident'. The Yokohama Doll Museum, one of Japan's largest, now exhibits some 1,300 models where wealthy Victorian travellers once took their tea and laid their heads. The mass of concrete, steel and glass angles pointing skyward is as contemporary as its predecessor, the Grand Hotel, was classic. A 'United Nations' of dolls, the museum believes it has a role in international relations. 'When you look at dolls from various countries, you can feel close to people you have never met,' museum curator Saeko Li says. The museum draws day-trippers and cruise passengers with time to kill before their ships depart.

Just a five-minute walk away, the Hotel New Grand was built in 1927 to replace the original. It is no longer new, but it is still very grand. The city fathers envisioned the New Grand as a symbol of Yokohama's rebirth after the disastrous Kanto earthquake. Inside I feel the pride and witness the excellence that has earned its reputation as one of the oldest and best western-style hotels in Japan. It speaks of the traditional Arts and Crafts movement prevalent in Nellie's day, together with a dash of art nouveau. The hotel's palatial stonework and abundance of earth tone tiles – *à la* American architect Frank Lloyd Wright – lend nobility to the elaborate interior accessed by a monumental staircase. Here my favourite architectural styles mingle in one legendary hotel that has hosted the likes of Charlie Chaplin, Babe Ruth, Jean Cocteau and General Douglas MacArthur. It feels a fitting successor to its celebrated forebear.

Outside the glazed entrance of the Hotel New Grand, an emerald ribbon of lawns and gardens skirts the waterfront for almost half a mile, uniting the city with its port. Like the hotel that fronts it, Yamashita Park symbolises the rebirth of the once-decimated city; perhaps even more so because this grassy esplanade rose directly from the rubble and ruins of the earthquake to become Japan's first seaside park. As I explore the grounds, the morning's drizzle escalates, forming mini puddles on the pavements, dousing the grass and seeping into the soil. I can smell the earth. This soft September shower does not discourage park visitors – spirited toddlers on tricycles and training bikes, their patient parents standing by, joggers, walkers, and sightseers like me trying to keep our cameras dry.

A short stroll away to the coast, ten concrete piers extend into the harbour like fingers, forming the port where Tokyo Bay meets the Pacific Ocean. The oldest, Osanbashi Pier, is constructed on the stone wharf where Nellie disembarked on 2 January 1890 after six days at sea. Today it welcomes the world's largest cruise ships. Emerging from the bay like a dolphin, Osanbashi unfurls in a colossal wave of timber, steel and concrete. Rolling wooden decking interspersed with grassy areas draws visitors to the pier's rooftop observation deck that locals call *kujira no onaka*, whale's tummy. As I gaze out on the skyline here, the scent of the sea mingles with the passing afternoon shower. In the distance, rising from the port's former shipyards, Yokohama's new futuristic city centre, Minato Mirai 21, dominates the horizon. The light rain paints a dreamlike gloss on this cluster of superstructures ready to supplant what little is left of Nellie's Yokohama.

Enough of the future, I am here for the past. Only yesterday's Yokohama can offer any trace of Nellie's journey. I wander away from the waterfront into the heart of old Yokohama where precious little pre-dates the earthquake. Seeking to enter Nellie's world, I follow Bashamichi Street, named after the horse carriages favoured by foreigners in her day, to its intersection with Minami Nakadori Street. There I climb seven stone steps, pass beneath a Romanesque arch and unlatch the century-old timber doors that lead into the Kanagawa Prefectural Museum of Cultural History. Time marches backwards in this Renaissance-style former bank with more than its fair share of Corinthian pilasters. Crowned by a green copper dome, this stately, but robust, building defied the earthquake. Alas, with a construction date of 1904, I know that Nellie did not see it. I have come to the museum to learn what she did see.

With the exception of a few courtesies, I cannot speak Japanese, and having a go in English and French is not helping me to explain where on the historical timeline I need to be. Museum steward Yuki Saito is here to guide me. Employing gestures and smiles, we somehow get there. Yuki, a young scholar whose brass-rimmed glasses frame grinning eyes, leads me into a library furnished in the same minty green as the copper dome looming outside. From rows of wooden shelving she pulls out several ledgers the size of laptop computers. Carefully labelled and positioned on manilla sheets, hundreds of hand-tinted postcards of old Yokohama are assembled three to a page. They are just the sort Nellie would have sent home to her mother. At last I am here, at the end of the nineteenth century, before the Great Kanto Earthquake. Inside these ledgers the city that Nellie experienced comes alive in individual 3 x 5 inch scenes. I capture Nellie's Yokohama by photographing postcards from the past.

The Grand Hotel returns in all its glory – its multi-gabled wooden façade festooned in flags and lanterns as cone-capped rickshaw drivers await their passengers. Best of all is 'Tea Service at the Grand Hotel'. Seven kimono-clad waitresses with lacquered hair pose behind a table of bamboo-handled tea pots ready to be served in the grand dining room. There is no sign of rats, but Nellie and I know they are there. 'View from Camp Hill' looks down on the port from a broad earthen lane lined by native houses. Nellie followed it to get to the 'Hundred Steps', also featured in many postcards, that led straight up to a shaded tea house with splendid views over the Yokohama that I am trying so hard to find. I am grateful for the time travel Yuki has arranged. There is another museum nearby that could shed some light, she indicates, handing me a leaflet for the Yokohama Archives of History.

Dashing back along the waterfront, I reach the Archives just as the doors are closing; the last entry was thirty minutes ago. Standing in the Archives courtyard, I discover that this building once housed the British Consulate, which was built in 1931 on the site where the Treaty of Peace and Amity was signed. Inside, the historical archives chronicle Yokohama's growth from a coastal settlement to an international port, beginning with the arrival of Commodore Perry. Nellie's trail in Yokohama still remains a mystery, but, here in the courtyard, I find myself following in the footsteps of US Commodore Matthew Calbraith Perry. I am too late for the exhibits inside, but there is some astonishing evidence outside. Rooted in the courtyard is the camphor tree that is said to have witnessed the signing of the treaty 160 years earlier, the treaty that opened the doors for Nellie's visit to Japan – and mine. The camphor is depicted in a watercolour painting entitled *Commodore Perry Coming Ashore at Yokohama* by the expedition's illustrator Wilhelm Heine. With branches reaching towards the bay where Perry's battleships anchor, the camphor stands tall and windswept on the right-hand side of the painting. This historic watercolour has been reproduced on a plaque for public viewing just a few feet from where the real camphor grows. Commodore Perry definitely beheld the tree, and so I believe did Nellie Bly. Now I have witnessed this mighty camphor, survivor of numerous fires, the Great Kanto Earthquake and the bombardment of Yokohama during the Second World War.

Along with the ancient camphor, the courtyard holds another reminder of Nellie's era. Posed behind the remarkable tree, an ornate lion's head water hydrant is a single representative of the multitudes that lined the city's main streets when Nellie was here. They were installed at 300-yard intervals in 1887, part and parcel of the country's first modern piped waterworks, the most technically advanced in the world at the time. The hydrants were above-ground portals for extensive underground iron pipelines spanning the countryside. Pure, clean water, free from the salt of the sea, streamed from the jaws of the lions and helped to curb the cholera epidemics that plagued the city. The waterworks were designed by British Major General Henry Spencer Palmer, who was also responsible for Yokohama's harbour and Osanbashi Pier.

The lion's head hydrant and revered camphor tree are the only tangible relics I find of a bygone era when Yokohama was young and so was Nellie as she made her way around the world, port by port. While she awaited the departure of the *Oceanic* for San Francisco, Nellie ventured into Tokyo and Kamakura for some sightseeing. So do I.

Chapter 11

In Which Nellie Meets the Great Buddha

The Great Buddha of Kamakura

Tokyo and Kamakura
2–7 January 1890

'My feverish eagerness to be off again on my race around the world was strongly mingled with regret at leaving such charming friends and such a lovely land.'

Nellie Bly

Like Nellie, it is my first visit to Japan. Seeking an authentic Japanese experience in Tokyo, I am booked into a *royokan*, a traditional inn, in the city centre. A *royokan* is the equivalent of a British bed and breakfast, only there is no bed and no breakfast. It is simply a room with a mat. Life takes place entirely on the floor. From a sunken dark panelled hallway with highly polished floors, I enter my 10 × 12 foot room through sliding paper doors called *shoji*. Inside there is a *tatami* mat made of woven rush grass and rice straw, and a futon, but not as we know them. A Japanese futon is like a malnourished duvet, three to four inches thick and packed with cotton batting. You sleep on it, not under it, atop your *tatami* mat. Every morning you roll up your futon and put it away to clear space for other activities in your room.

Nellie fared better. She slept in a real bed with a proper mattress at the Grand Hotel in Yokohama. Her '*royokan* experience' occurred when she joined other guests to visit a Japanese home for a performance by Geisha girls. As soon as you enter a home or a *royokan*, you must remove your shoes and don the mandatory cloth slippers provided by your host. Nellie's group of foreigners were having none of that. 'At the door we saw all the wooden shoes of the household, and we were asked to take off our shoes before entering, a proceeding rather disliked by some of the party, who refused absolutely to do as requested. We effected a compromise, however, by putting cloth slippers over our shoes.' The group then entered a large room, empty except for a *tatami* mat and a few Japanese screens. 'We sat upon the floor, for chairs there are none in Japan, but the exquisite matting is padded until it is as soft as velvet. It was laughable to see us trying to sit down, and yet more so to see us endeavour to find a posture of ease for our limbs. We were about as graceful as an elephant dancing,' Nellie wrote.

The *tatami* mat in my *royokan* is more like sackcloth than velvet. It prickles my feet. Like Nellie, I never really discover a 'posture of ease' on the floor. I have it easy compared to my Victorian predecessor with her floor-length broadcloth gown. In my whisper-light travel garb, I can just flop down. But just the same, I make this my first and last night in a *royokan*. I now understand the importance of mattresses and chairs in my life. I opt instead for a high-rise hotel in the buzzy Shinjuku district with a queen bed, a desk, a chair, free Wi-Fi, toiletries, towel, and cloth slippers I am not obliged to wear. Adjacent to Shinjuku Station – the world's busiest railway station – my corporate hotel edges on Japan's largest and wildest red-light district, where gaudy neon-encased towers blaze and flash, luring passers-by inside for cocktails, karaoke and adult entertainment. Compared to Amsterdam and Las Vegas, this area known as Kabukicho feels quite tame to me. Once I am checked in to the hotel, I depart from the seedy side of the city towards the majestic. Thirty minutes away by subway, the Tokyo Imperial Palace, official home of the Emperor, is my first stop on Nellie's trail here.

Like an island in a sea of skyscrapers, the emperor's 'royal realm' is a haven of parks, gardens and woodlands. It rises from the ruins of the fifteenth-century Edo Castle, headquarters of the feudal government until 1867 when Emperor Meiji became Japan's ruler. On the throne when Nellie was here, Meiji was the first monarch to live, work and govern at this location. Four successions later, the newly crowned Emperor Naruhito now resides at the Imperial Palace. When Nellie visited the royal residence on a day trip from Yokohama, she was welcomed into the inner grounds. I am not. But I get a good view from the plaza of the Kokyogaien National Gardens that front it. Now bequeathed to the public, these gardens formed part of the royal grounds until 1949. From here I can see large expanses of the palace's private property – a guardhouse, a gate and, in the distance, the famous watchtower with its grey winged roofs. Squatting on a massive stone remnant from the Edo Castle, the three-storey watchtower is encircled by a forest of Japanese black pines. Their wavy silhouettes and frizzy canopies dominate the grounds here. I photograph the pines against the watchtower; in the foreground, an elegant double-arched bridge reaches across the moat. Without realising it, I have captured on camera one of Tokyo's most iconic views – the Fushimi-yagura watchtower and the Nijubashi Bridge, gateway to the Imperial Palace.

Northwards across Tokyo, I head for Japan's first and most popular city park – Ueno. The Japanese counterpart to London's Exhibition Road in South Kensington and Berlin's Museum Island, it hosts an impressive collection of world-class museums including the Tokyo National Museum, Tokyo Metropolitan Art Museum and the National Science Museum. Ueno Park sits in the grounds of the seventeenth-century Kaneiji Temple which shares its lands with the museums, carnival rides and Japan's first zoo.

Recounting her visit to Ueno Park, Nellie mentions only a specific tree and a frisky monkey she encountered in the zoo. It was the monkey that grabbed her attention. 'At Uyeno Park, where they point out a tree planted by General Grant when on his tour around the world, I saw a most amusing monkey which belonged to the very interesting menagerie.' I opt out of a zoo visit and go in search of the tree planted by the American General who served as the eighteenth President of the United States when Nellie was a girl. Some 8,800 trees grow at Ueno. With help from the park keepers, I am able to track down not one, but two trees, planted by General Ulysses S. Grant and his wife Julia on the world tour they began in May 1877, two months after he left the White House. Like Nellie before she set off on her global journey, the Grants were exhausted and felt in need of a vacation. Eight years of stitching together a nation torn and bleeding from the Civil War had taken their toll. Their journey pre-dated Nellie's by a dozen years; but the General and the former first lady were not racing. When they departed from the port of Yokohama to return home on 3 September 1879, the Grants had been away from America for two and a half years. During his international travels, the former US president had helped to place America firmly on the map as a global player. His legacy in Tokyo is commemorated by the two trees and an open-air memorial that enshrines them. The trees emerge from a sarcophagus-like planter embedded with a copper triptych featuring the General's likeness and excerpts from his speeches. It forms the centrepiece of the memorial enclosed on both sides with built-in concrete benches. As I explore the site, toddlers in matching red t-shirts scramble on one bench while their parents relax on the other. Children queue for ice creams nearby as shrieking brakes and repeated collisions sound from the neighbouring bumper-car rink where drivers race and swerve, oblivious to the venerable trees that have drawn me here. It strikes me that Nellie would have been of the 'dodgems' age when the Grants made their journey.

Where their branches meet in the sky, the lustrous leaves of the magnolia tree planted by Julia Grant provide a foil for the dense, dark foliage of her husband's cypress. Both are evergreens, symbols of perpetuity and survival. Earthquakes, bombs and fires have failed to topple these time-honoured trees since they were planted 140 years ago as 12-inch saplings. I cannot be sure what General Grant said as he and Julia placed their seedlings in the earth, but perhaps it was comparable to the words he spoke in Nagasaki at an earlier tree planting. 'I hope that both trees may prosper, grow large, live long, and in their growth, prosperity and long life be emblematic of the future of Japan.'[1] These were prophetic words for Tokyo, if not for Nagasaki.

Unlike Nellie, who was passing through, General Grant was an official guest of the nation. Ceremonies and receptions awaited him at every stop on his Japanese tour. But the greatest of all, a *Fête Champêtre*, took place at Ueno Park on 25 August 1879.[2] The tree planting ceremony that afternoon was among the more solemn of the day's festivities in honour of the General. The celebrations drew the country's finest archers, horsemen, marksmen and fencers to perform before the Emperor, his American guests and selected citizens of Tokyo. Fireworks blazed across Tokyo's night sky during the dazzling *grand finale*, painting the darkness in diamonds and rainbows. Whistles, sizzles and blasts ricocheted across the park as Roman candles and Catherine wheels exploded. Wispy streams of smoke dangled like ghosts in the air before dissipating into the summer evening. A week after the *Fête Champêtre*, before boarding the SS *City of Tokio* in Yokohama for his return to America, General Grant declared his visit to Japan the 'most pleasant' of all of his travels. 'The country is beautifully cultivated, the scenery is grand, and the people, from the highest to the lowest, the most kindly and the most cleanly in the world,'[3] he said. Just like Nellie, General Grant was enthralled by Japan, calling it 'beautiful beyond description'.

The World's globe-girdler was not following in General Grant's footsteps, and neither am I. But our paths converge again at Zojoji Temple in Shiba Park, forty minutes south of Ueno by train. It was known as the Great Shiba Temple when Nellie and the Grants were here. Today's Zojoji, one of Japan's principal Buddhist temples, stands in the heart of Tokyo as a hub for events as zany as soybean throwing festivals, and as traditional as masked Noh theatrical performances dating from the fourteenth century. Nellie and the Grants would immediately recognise the majestic entrance to the temple grounds, the imposing Sangedatsumon Gate, which was constructed in 1622. Rising two stories and stretching sixty-nine feet long, Sangedatsumon's age and eminence are evident in its scale, its patina and its presence. It is the oldest wooden structure in a country prone to fire and earthquakes. Just inside this ancient main gate stands a Himalayan cedar planted by General Grant two weeks before the cypress tree ceremony at Ueno Park. Ascending into a cloudless autumn sky, the cedar's emerald crown is set against the burnished vermillion of the revered gate. Both have endured all that nature and humanity have flung at them – natural disasters, war and so-called progress. Spiritually, the Sangedatsumon Gate is said to mark the transition between the mundane and the sacred. Once past the gate, we are delivered from three earthly states of mind – greed, anger and stupidity. Today as I enter the temple grounds, I am transported from the commotion of one of the world's largest cities into the aura of calm that exists on the other side of Sangedatsumon Gate.

But not for long. As I make my way towards the temple's main hall, the Daiden, hundreds of people are spilling out of tour buses. They are not tourists; they are worshippers who have come to Zojoji to pay homage to the Amida Buddha, the Buddha of Limitless Light. In the grounds behind the Daiden hall, the Mausoleum of the Tokugawa Shoguns holds the remains of six of the fifteen military rulers of Japan's Edo period. It is splendidly built in the style of the times with graceful sloping roofs, ornate carvings, lacquer and gold leaf ornamentation. But I am more drawn to an outlying garden, where rows of identical stone statues of children stand in silence. There are more than one thousand. Posed in prayer with tiny folded hands, upturned faces and soft smiles, these simple statues represent babies that were never born due to miscarriage, abortion or stillbirth. It is the Garden of the Unborn Children. About eighteen inches tall, the statues are called *Jizo*, the protectors of babies who did not have time to acquire the karma on Earth that will ease them into the afterlife. Mourning Buddhist parents adorn their *Jizo* in red crocheted hats; many have pilgrim-like collars or bibs worn over their stone-carved robes. Vibrant silk sunflowers, roses, daisies and lilies sprout from pale grey vases adorned with calligraphy and placed before each statue. Next to the vases, plastic pinwheels in crayon-box colours whirl in the autumn currents, helping to lift the prevailing solemnity. I am here alone. I did not expect to see this memorial garden; Nellie certainly did not. But as I regard the statues before me, I am reminded of the crusades she launched throughout her journalism career for vulnerable children and parents. Although she had no offspring of her own, Nellie impacted the lives of many children and families. Her newspaper reports raised funds to treat sick children, rallied support for mothers-at-risk and found stable homes for abandoned babies. Here at Zojoji 'cemetery' for unborn children, it is the parents who have been deserted.

* * *

With the boxes ticked on my list of places that Nellie saw in Yokohama and Tokyo, it is time to swap urban for rural Japan and follow her to the coastal city of Kamakura, where shrines and temples flourish in the forested mountains and valleys nearby. Kamakura lies forty miles south of Tokyo and sixteen miles from Yokohama. Nellie travelled there from Yokohama. I travel from Tokyo with two Japanese friends.

'I went to Kamakura to see the great bronze god, the image of Buddha, familiarly called Diabutsu,' she wrote. Nellie did not realise that although he is holy, Diabutsu is not a god. 'It stands in a verdant valley at the foot of two mountains.' The 'verdant valley' where Diabutsu reigns now shares its ancient lands with barista-staffed coffee bars, organic ice cream parlours, and souvenir shops selling designer chopsticks and seashell jewellery. One of Tokyo's top day trip destinations, Kamakura now attracts some two million tourists a year.

I am among them thanks to the kind invitation of my Japanese friends, Yoshihisa and Yoshie Togo. The children's charity UNICEF first brought us together. Yoshihisa led Japan's Committee for UNICEF at the same time that my husband, David Stanton, was heading up UNICEF-UK. The cause united us, and an East-West friendship was born. Yoshihisa has devoted more than thirty years to raising funds for children in danger, leading the Japanese to become one of the world's major donors to UNICEF. He is a significant international player, but you might not know it by his modesty, openness and winning smile. Yoshie is equally cosmopolitan and ever so gracious.

When I know my itinerary will include Japan, I let the Togos know that Nellie Bly had fallen in love with their country and I would soon be following in her footsteps. They are pleased to learn that I will be raising money for UNICEF through my travels, in honour of Nellie and her work on behalf of children. Knowing their hectic schedules, I expect no more than a hurried lunch or dinner together in Tokyo. Instead, Yoshihisa and Yoshie invite me to join them for an overnight journey into the Japanese countryside. I jump at the chance to see them again and experience another side of Japan, even if it means an unplanned break from my unrelenting Nellie Bly quest. But that is not to be the case. With no prior knowledge of Nellie's itinerary in Japan, the Togos propose a night in Hakone, their favourite mountain get-away, with a visit to Kamakura the following day. They are going to deliver me to the very place where Nellie encountered the Great Buddha. This is another of those serendipitous twists of fate that enlighten my travels in the name of Nellie Bly, and let me know I am on the right track. The Togos did not know that she had been there before them; indeed, like everyone else I meet, they have never heard of Nellie Bly. Why should they? The purpose of my trip is to get her back on the map by recreating her journey. I feel she has been forgotten, consigned to obscurity at a time when we could do with more trailblazing female role models. Little did I expect others to take on my mission with the kindness and enthusiasm shown by the Yoshihisa and Yoshie, and indeed others, along the way.

By the time I meet the Togos for our foray into the Japanese outdoors, I have travelled half-way around the world. I have spent thirteen days and nights constantly pursuing Nellie's epic journey across seven countries. I have gone for many days without a meaningful conversation with anyone except myself; and an occasional chat to Nellie. I have never felt alone, but I am now more than ready now to join friends to discover the country that she so adored.

We agree to meet outside the Shin-Yurigaoka Rail Station, southwest of Tokyo, where Yoshihisa and Yoshie will pick me up in their car. It is a Sunday morning. As my train from the city centre glides along the platform, this normally heaving station exudes a weekend sense of calm. I am glad for the absence of commuters as I make my way to the exit. I have arrived twenty-five minutes early for fear of being late. Apprehension is mounting as I think about inflicting myself on this gentle couple for two whole days.

Spot on time at 8.45 am, a deluxe white sedan pulls up and I slide into the spacious back seat upholstered in leather. My apprehension dissolves as soon as I enter the car. No need for an icebreaker. We launch into chatter that sets us all at ease immediately. They ask first about David, who they know better than me. 'He's fine, immersed in UNICEF and helping our daughter Acadia to prepare for her second year at university,' I report. I tell them how supportive David has been of my Nellie Bly travels, and that it is he who suggested that I contact them about my trip. 'But now I am here, he is slightly jealous not to be with us,' I say. As devoted anglophiles, Yoshihisa and Yoshie are well-versed on all things British. We review the forthcoming elections, the Commonwealth

Games and the latest royal news. I am keen to tell them of my discoveries in Tokyo and Yokohama, and how, just like Nellie Bly and General Grant, I am enthralled with their country. That is even before I have seen the enchanting landscapes they will soon introduce to me.

We are on our way to Hakone, the land of volcanic hot springs, mountains, forests and lakes. With their conviviality and international charm, the Togos are the consummate hosts. For the next forty-eight hours I will be in their friendly and capable hands. No constant searching for sites, no tortuous map reading, no navigating subways and buses, no trying to bridge the language gap. I write later in an email to David: 'I am relaxing for the first time since I left home because I don't have to try to find anything. Yoshihisa and Yoshie are making all of the arrangements and all I have to do is enjoy them.'

The Togos are dressed in sand-coloured khakis and blue shirts – his is a sky-blue polo and hers a narrow-striped navy pullover. Since our previous contact has revolved around UNICEF meetings, I have only known them in business attire. It is nice to see them in casual mode. Their soft summer tans have not faded. Driving through Hakone National Park, we stop at viewpoints along the way, posing for each other's cameras, and savouring the natural beauty that embraces us on this balmy September day. At Lake Ashi, a neon blue crater lake sprinkled with pleasure boats, Yoshihisa and Yoshie slip behind a life-size cartoon cut-out board that converts them into a Samurai warrior and a Geisha girl. Only their faces show in this lakeside photo opportunity staged for Instagram fans. I seize these moments digitally. Some of the most exquisite views of Mount Fuji are found here. I think back to London and the print of David Hockney's *Mount Fuji and Flowers* that hangs in our hallway. He painted it from a postcard. With my Sony Cybershot, I attempt to capture the drama of Fuji's exceptionally symmetrical cone emerging from a sea of clouds rising above sylvan ridges. This archetypal peak shares the stage with Italy's Vesuvius as the world's best-known volcano.

Our hotel, the Fujiya, is named after the volcano that is also Japan's highest peak at 12,390 feet. With a symphony of pagoda roofs outside and heritage furnishings inside, the Fujiya's exotic blend of east and west would have been a style familiar to Nellie. Since its opening in 1891, this historic hotel has hosted, among others, Charlie Chaplin, Helen Keller, John Lennon,

Yoko Ono and royalty from around the world. The guest rooms are set in lush gardens where pathways meander past stone-sculpted lions and temple lanterns, and down rocky steps to ruby red bridges crossing rock-strewn pools. A thatched watermill and its revolving wheel draw honeymooning couples pond-side for scenic photographs. Not far away, the Fujiya's vine-cloaked waterfall spills into a natural basin where carp as large as your hand swim in circles. Inside, the dark wood-panelled lobby features a massive winding staircase carpeted in red. Uniformed bellhops with pillbox hats enhance the old-world elegance of this iconic hotel celebrated for its under-stated grandeur and *onsens* – natural hot spring baths.

With more than a dozen hot springs bubbling to the surface, Hakone is a Mecca for *onsen* bath lovers. At the Fujiya alone, I can dip into three private *onsens*, a group *onsen*, as well as indoor and outdoor hot spring pools. This hot mineral water is even piped into my own bathtub. I try them all, with the exception of the private baths. Hygiene is key. Before bathing, you must sit on a stool and scrub yourself thoroughly with a brush. After weeks of fast-dash showers, I feel squeaky clean after a series of before-dinner soaks in the hotel's assorted *onsens*.

Unlike Nellie, I was able to squeeze a few garments for special occasions in my bag. Glad to swap chinos and t-shirt for a silk skirt and ballet pumps, I dress for dinner as one does at a posh period hotel, especially if you are dining with the Togos in a classic French restaurant with silver service. Inside this lavish dining room with dragon flourishes, we are seated beneath a coffered ceiling painted with pretty alpine flora and fauna motifs. Hilltop scenes outside spill inside through the abundance of windows around us. After a series of in-room and park bench picnics, the crisp damask tablecloth and matching serviettes seem almost too lush to use. I indulge in cream-laced *Coquille St Jacques* with equally sinful *gratin Dauphinoise* served on china decorated with hand-painted images of Mount Fuji. My *Chenin blanc* is decanted into a crystal glass by a white-gloved *sommelier* who waits while I admire its pale honey colour, draw in its floral aroma, and sample the hints of apple and pear it presents. *A votre santé* Yoshie and Yoshihisa. My first break from Nellie in two weeks is a welcome one. I will catch up with her tomorrow when we go to Kamakura.

* * *

It is Monday afternoon and we are standing before Diabutsu, the Great Buddha of Kamakura. Two women in angel-sleeved kimonos are in front of us – one in black silk sprinkled with tiny pink blossoms; the other in peach linen embossed with vanilla maple leaves. Their glossy black hair is woven into tight chignons and garnished with fresh flowers. We are all witnessing one of Japan's greatest icons. Diabutsu commands the entire region. Behind him, tree-carpeted hills roll into a powder-blue horizon where clouds rise like smoke signals. Beside him, sacred bronze lotus flowers curl skywards towards his shoulders. This monumental Amida Buddha is bounded by filigreed lanterns set on pedestals that stand on both sides like sentries. Incense trails drift over and linger from the adjacent temple of Kotoku-in. In the far distance, to the west, Mount Fuji is watching. Everything feels so wonderfully Japanese and timeless. Except for the daypack-toting tourists taking selfies with Diabutsu, we could be here with Nellie in 1890. I can see her detailed descriptions of the Great Buddha come to life.

She must have taken notes from a guidebook, for Nellie reports in minute detail all of Diabutsu's vital statistics from his soaring five-storey height to the circumference of his thumbs. '[F]ifty feet in height; it is sitting Japanese style, ninety-eight feet being its waist circumference; the face is eight feet long, the eye is four feet, the ear six feet six and one-half inches, the nose three feet eight and one-half inches, the mouth is three feet two and one-half inches, the diameter of the lap is thirty-six feet, and the circumference of the thumb is over three feet,' she wrote. 'I had my photograph taken sitting on its thumb with two friends, one of whom offered $50,000 for the god.'

I would love to find that photo; indeed any photo of Nellie on her travels. To get a sense of what she experienced, I make do with an internet-sourced 1892 image of three men in top hats, two carrying canes, smiling for the camera while seated on the hands of the Buddha. In their day, wooden steps led up into the Diabutsu's lap. His long, slender fingers are folded to create a platform for his thumbs in the Buddhist *mudra* (gesture) that allows for the highest level of enlightenment. I must admire his sofa-sized thumbs from a distance; but when Nellie was here, as many as six people could pose on them comfortably. Today a Buddha-sized bowl of fresh fruit and a colossal spray of autumn flowers lie at his enormous fingertips.

Back then, the Great Buddha was as accessible on the inside as he was on the outside: 'the hollow interior is harmlessly fitted up with tiny altars and a ladder stairway by which visitors can climb up into Diabutsu's eye, and from that height view the surrounding lovely country,' wrote Nellie of her ascent into the holy head. Climbing up to the Buddha's eyes is no longer an option, but his belly is still open for viewing. Yoshihisa and Yoshie have visited Diabutsu before, so they leave me to experience this phenomenon alone. I am thrilled to think that, like Nellie, I can step right inside this Buddha. I never imagined that all I would need is a twenty yen (fourteen pence) ticket to be so close to Diabutsu, and indeed to Nellie. Narrow steel stairs lead into his vast paunch, which is as dark as a cavern and smells of metal. Illuminated drawings demonstrate the remarkable construction in the mid-thirteenth century of this massive bronze Buddha cast in forty separate sheets, brazed together, and finished off with a chisel. Squinting, I can detect some of the patchwork-like seams. Since his creation, Japan's second largest Buddha has survived wars, natural disasters and homelessness. The Temple Halls that once lodged him burnt down more than once, so Diabutsu has resided in the open air since the fifteenth century. Bombs and earthquakes have barely grazed him; even Japan's most powerful quake, the Great Kanto, only managed to budge him a couple of feet from his Great Buddha base, as I learned in Yokohama. And now I am here, 125 years after Nellie. There are only two places in my entire journey where I can be certain that I am standing exactly where Nellie stood; two locations where I know that my footsteps are her footsteps. The first was in Jules Verne's home in Amiens, where I shared his salon, study and winter garden with Nellie. The second is here in the belly of the Great Buddha of Kamakura.

My close encounter with the Buddha, and with Nellie, is a fitting finale to a two-day journey that feels like it will last much longer. It is time to return to Tokyo. My horizons have expanded, and so has my friendship with the Togos. This makes it especially difficult to say goodbye when they drop me off at the station, the site of our rendezvous just the day before. I return rather solemnly to a Tokyo slipping into dusk.

Nellie often lamented her departures, despite her burning desire to race on. Although it was a 'bright sunny morning' when she left Yokohama on Tuesday 7 January 1890, Nellie's heart was torn by mixed emotions.

'My feverish eagerness to be off again on my race around the world was strongly mingled with regret at leaving such charming friends and such a lovely land,' she wrote. But what a send-off she received. 'A number of new friends in launches escorted me to the *Oceanic*, and when we hoisted anchor the steam launches blew loud blasts upon their whistles in farewell to me,' she wrote. Also ready to see Nellie off in style was the SS *Omaha*, an American sloop anchored in Tokyo Bay. The captain and crew had recently hosted a luncheon on board in honour of Nellie, 'one of the pleasant events' of her stay. As the *Oceanic* sailed through the peaceful waters of the new port into the vast and daunting Pacific Ocean, the naval band of the SS *Omaha* serenaded Nellie with renditions of 'Home, Sweet Home', 'Hail Columbia', and 'The Girl I Left Behind Me'. 'I waved my handkerchief so long after they were out of sight that my arms were sore for days,' Nellie wrote.

Nellie sailed from Yokohama across the Pacific Ocean to San Francisco, from where she travelled cross-country by train to New York City, which is where I will join her.

Chapter 12

In Which Nellie Triumphs

Jersey City railway and ferry terminus

Hoboken and Jersey City, New Jersey; New York City
14 November 1889, 25 January 1890

'I took off my cap and wanted to yell with the crowd, not because I had gone around the world in seventy-two days, but because I was home again.'

Nellie Bly

Nellie swapped steamships for steam trains for the final phase of her journey back to New York City. She was on the 'home stretch' in her home country. Arriving in San Francisco after a fourteen-day passage on the SS *Oceanic* from Japan, she was now 'flying' southeast across the country on a train chartered by *The World* to bypass impenetrable snowdrifts further north. 'I only remember my trip across the continent as one maze of happy greetings, happy wishes, congratulations, telegrams, fruit, flowers, loud cheers, wild hurrahs, rapid hand-shaking and a beautiful car filled with fragrant flowers attached to a swift engine that was tearing like mad through flower-dotted valley and over snow-tipped mountain on-on-on!' she wrote. 'It was glorious! A ride worthy of a Queen.'

Multitudes of well-wishers dressed in their Sunday best packed train stations on the route to cheer Nellie on as she traversed the nation on a whistle-stop tour – Fresno, Albuquerque, Topeka, Dodge City, Kansas City, Chicago, Columbus, Harrisburg. In Pittsburgh, where her journalism career began, thousands turned up at Union Station at 3.10 in the morning to greet their Nellie. With tears in her eyes, she stepped onto the rear platform of her railway car and waved to all those who had left their beds in the middle of the night just to see her. Despite her celebrity status, their globe-girdler had not changed, said *The Pittsburgh Commercial Gazette*. 'She was the same bright-eyed girl who used to hustle … in this city, with a cheery greeting for every friend and an eager desire to gather in any stray gems of thought…' the paper reported. When she appeared on the platform, she looked the 'picture of health'. 'I am feeling splendid,' Nellie told the *Gazette*'s reporter. 'I am not in the least fatigued and have had good luck during my entire trip.' Her next stop in Philadelphia, 300 miles east of Pittsburgh, would be the final one before Jersey City, the official finish of her race. Nellie was just twelve hours away from her home in New York City … and the end of her triumphant world journey.

*   *   *

It is 6.30 pm and I am seated in seat 48C for take-off on Japan Airlines flight JL4 from Tokyo's Narita Airport to John F Kennedy (JFK) in New York City to catch up with Nellie. It is the fifth and longest lap of my round-the-world itinerary. At thirteen hours, it is also the longest flight I have ever taken. I am leaving Asia after eighteen days on the trot tracing my predecessor's footsteps steps across Ceylon, Singapore, Hong Kong, China and Japan. The melancholy I could feel about leaving this exotic continent subsides with thoughts of the people I will see in America, starting with my lifelong friend Alice Robbins-Fox who is joining the Nellie Bly trail in New York City. So far I have travelled solo just like Nellie, but I think I can justify a companion once I reach New York.

We depart from Tokyo on schedule. Thanks to the international dateline, I will arrive in New York City at the same time, on the same day – Wednesday 24 September – that I leave Tokyo. Travelling this way around the world, I have gained a day. I like this phenomenon. It is the *denouement* that allowed Phileas Fogg to complete his world journey in eighty days, win his wager and collect £20,000 (more than £2 million/$2.5 million today) from his friends at the Reform Club in London. Across the world in another time zone, Alice will depart at 3.45 pm on Jet Blue flight 1784 from Florida's Orlando International Airport. With a little research and a lot of luck, we found flights that arrive at JFK within seven minutes of each other.

My lengthy flight seems to literally fly by. Perhaps because I am reclining in one of the world's best economy class airline seats known as the 'Sky Wider' for its generous width and sleek pitch. The scent of spicy prawns wafts from my fresh seafood dinner that also includes salmon sushi, squid *sashimi* style and sticky rice that I consume using thin wooden chopsticks just like Nellie described. 'At a tea-house or at an ordinary dinner a long paper laid at one's place contains a pair of chopsticks, probably twelve inches in length, but no thicker than the thinner size of lead pencils. The sticks are usually whittled in one piece and split only half apart to prove that they have never been used. Every one breaks the sticks apart before eating, and after the meal they are destroyed,' Nellie explained to her readers. She admired the 'proper and graceful way' others employed them: 'It is a pretty sight to see a lovely woman use chopsticks,' she wrote. With the last of my prawns gripped in my chopsticks, I ask for another glass of the airline's oaky Chardonnay to sip while I watch *The Grand Budapest Hotel* on the in-flight entertainment channel. The vintage fictional hotel in this 2014 comedy-drama film takes me back to my stays amid the faded glory of the Grand Oriental Hotel in Colombo and the classic splendour of the Fujiya in Hakone where bellhops still wear the same pillbox hats as the Grand Budapest's lobby boy Zero. At last I can afford to relax. On all the other flights I was landing in unknown territory with foreign languages and unfamiliar cultures, so I avoided alcohol and prepared myself mentally and physically to hit the ground running. But going to America is almost like going home. I have spent half of my life in the United States, growing up in Florida and living with David in Washington, DC during his posting there. I have British citizenship from my parents; I am Canadian by birth and American by naturalisation. Like many in search of their roots, I came to England, home to all of my relatives, in the 1980s and decided to stay. It is nice to have more than one passport – United Kingdom and USA – when you are travelling around the world. Nellie left America without one. She had to pick up her official passport at the American Legation in London. Now, with my two passports in hand, I am approaching the city where Nellie's journey began in trepidation and ended in triumph.

By the time my aircraft lands at JFK airport, I have circled more than a third of the world just on this flight. I disembark and dash off to immigration where everything is automated and the queues are short. All I have to do is scan my passport at a kiosk which also records my fingerprints and takes my photo. Zigzagging through the crowds flowing into the Terminal One arrivals hall, I position myself in a conspicuous spot and forage in my bag for my mobile phone. I am keying in Alice's number when I hear her voice. 'Is that you Nellie Bly?' Alice says with a half-moon smile. She is standing right beside me. How she got here and found me so fast I do not know; maybe it is because she is an emergency room nurse. As we hug and thank the travel gods for uniting us so quickly, I catch a glimpse of the massive American flag floating far above us. 'Welcome to the USA,' says Alice as we dash towards the AirTrain that will deliver us to the New York subway and into Manhattan. We sprint inside and grab overhead hand-straps for the five-minute journey. The AirTrain darts out of the terminal station into an apricot sky flocked in luminous pink and blue like a flaming opal; the horizon is an iridescent jewel. We are momentarily silenced by this unexpected gift. It is only thirty minutes since we landed at JFK and we are riding into a Manhattan sunset. We change onto subway line E at Jamaica station, then line F. Our conversations travel faster than the train. Alice listens readily between the screeching subway stops as I recount a few of my escapades. A native-born Floridian with amber hair, deep-seated wisdom and a dusting of freckles, Alice is well acquainted with hurricanes, but even so her chestnut eyes expand as I tell her about my run-in with the Asian equivalent, the relentless Typhoon Kalmaegi in Hong Kong. In forty years of friendship, we share a history that I treasure. Adjacent passengers, irritated by our spirited chatter, glare into their mobiles and boost the volume on their earphones. After spending so much time alone, I am oblivious to their annoyance, and so is Alice.

Darkness has tumbled on Manhattan as our subway train rattles into the station at Lexington Avenue and Sixty-Third Street on the Upper

East Side. Skyscrapers glitter above us and neon glares around us as we climb the station stairs and exit onto a broad sidewalk. The smoky-salt scent of grilling hotdogs floats over from a street vendor's cart as yellow cabs cruise by on their way to Park Avenue. Alice and I have room reservations at the nearby Cosmopolitan Club, courtesy of my city friend and club member Sally Emery. She is one of the accomplished females who rub shoulders at this private club established in 1909 for women to 'nourish their intellects, exercise their artistic impulses; cultivate friends; and freely exchange ideas'. Eleanor Roosevelt, Margaret Mead, Willa Cather, Pearl S. Buck, Marian Anderson and other leading lights in the arts, letters and public affairs were members. Elizabeth Jane Cochrane, aka Nellie Bly, was not, although the 'Cosmo' Club seems just the place for her. That aside, this Neoclassical-cum-Art Deco grey brick building famed for its wrought iron balconies is the ideal New York City headquarters for the Nellie Bly trail team – Alice, Sally and me. We will meet tomorrow morning in the Sunroom, 'the crown of the Cosmopolitan Club', on the tenth floor.

After breakfast Sally joins us at the club. Baltimore is her hometown, but this silver-blonde business executive has adopted New York as her own and knows it like a sister. Among other roles, Sally is a volunteer gardener in Central Park and an intrepid traveller like Nellie, exploring much of the world, especially Asia. This morning we have the Sunroom to ourselves. September sunshine streams in through the floor-to-ceiling windows, coating everything in a crystalline glaze. A wrap-around tiled balcony beckons us outside to greet the city. We do. Honks and hisses ascend from the streets as we step out onto this tenth-floor viewpoint. The Midtown Manhattan skyline rises before us in a medley of imposing rectangles – some stocky, most soaring, all vertical and crisscrossed with carbon copy windows. Amongst them we spot a handful of wooden rooftop water towers. Some 15,000 of these hand-built casks with legs and ladders are perched on tall buildings across Manhattan. They originated in Nellie's day as the city grew upwards and water pressure could reach no higher than six storeys. These towers still provide drinking water to New Yorkers, Sally explains. She will use that same New York City know-how to chart a trail for Alice and me to follow Nellie. Back inside she flattens a multi-fold city map onto the shiny floor beside diagrams of subway and bus routes. We gather round on our knees, tracing potential Nellie paths with our fingers. Sally has also arranged a visit to Nellie's final resting place at Woodlawn Cemetery in the Bronx. But first things first – Alice and I are sent straight down to lower Manhattan to re-enact Nellie's cautious departure from the docks at Hoboken, followed by her jubilant arrival at the rail terminus in Jersey City. They are both across the Hudson River in New Jersey. We will start and finish the race exactly where Nellie did, but in one day – today – rather than seventy-two. It is just a matter of fast-forwarding.

The weather was crisp and clear on the day Nellie boarded the SS *Augusta Victoria* at the Hamburg-American Packet Line docks in Hoboken. Not so for Alice and me. The mid-morning sun has surrendered to cement skies and patchy showers on our way to the waterfront to cross the Hudson. We arrive at a rain-drenched Pier Eleven near Wall Street, only to discover that the harbour is off-limits and all ferries are suspended. We must wait. Even Nellie 'don't take no for an answer' Bly would be obliged to cope with this delay. We are trying to discover the cause when two Marine Corps Osprey aircraft swoop in like pterodactyls, slashing the skies with their whooshing grey blades and hovering ominously over the Downtown Manhattan Heliport nearby. A pair of helicopters whisk in from the sides to join the panoply. We shield our ears from the metallic roar. Alice points up to three armed men – snipers – crouched on the roof of the heliport headquarters. A Coast Guard cutter plies the waters as police cars, blue lights flashing, form a barricade separating the waterfront from the city. We are on the wrong side. Rain streams down; umbrellas pop up. A motorcade of black limousines crawls like a funeral cortege onto the heliport pier where the Ospreys have now landed. A dozen people in black suits exit the limos and climb aboard the aircraft awaiting them. President Barack Obama and First Lady Michelle Obama are among them. They are returning to Washington, DC after three days of United Nations talks on climate change, education rights and the Ebola epidemic. The Ospreys rev up, propellers whirring and whining, and ascend. They are flying to JFK airport, where Alice and I arrived only last night, to board Air Force One. Within minutes of lift-off, the scene is cleared, the ferries are back in business, the Obamas are en route to the White House, and we are heading for Hoboken, where Nellie stepped off the docks and into history 125 years earlier.

Ships of all sorts shared Hoboken's harbour then – ironclads, brigs, tugs, paddle-steamers, clippers and ferries. Transatlantic liners heading for Europe and beyond extended like horizontal rocket ships from horse shoe-shaped berths. Forests of spindly masts were etched in the skyline, and townhouses spilled onto a waterfront teeming with passengers, crew, chandlers and merchants. Not now. The demise of the steamer routes brought the decline of the dockyards. These days only commuter ferries from Manhattan, such as the one that Alice and I have boarded, tie up at a Hoboken riverside now stripped of its maritime enterprise. Today the once-blighted waterfront has been re-invented as a trendy residential quarter with luxury apartments and swish boutiques and restaurants. After a blustery river crossing, Alice and I disembark our ferry at Hoboken Terminal, the jewel in the crown of the revived waterfront.

Inside this historic depot, the last working survivor of the five great Hudson River railroad terminals, we escape the persistent drizzle. Decorative copper exterior façades host Beaux-Arts interiors, lit by an enormous Tiffany glass skylight framed in tiers of sculpted plaster. Passengers relax in banks of back-to-back wooden benches capped with butterscotch library lamps in this concourse that epitomises the golden age of American rail travel. Cameras in hand, we wander around capturing architectural gems such as the fluted chandeliers that dangle like earrings from soaring ceilings, the citadel-style wall torches and a bronze lion-head ice water fountain. A Victorian-style clock tower outside oversees this lavish 1907 transport depot listed on the National Register of Historic Places. It is all very beautiful, but Nellie is not here. This is not the Hoboken my predecessor saw the day she departed. It is hard to capture the mood; we will have to leave it to the journalists at the scene on that auspicious day.

'Upon deck Miss Bly found plenty of friends wanting to shake her hand for a last good-bye before her long and yet short journey – long in miles and short in time,' *The New York World* reported. 'A finer morning for a start on a sea trip could not have been chosen, and the crisp November air freshened her fair young cheeks as she stood blushingly in the center of a group of admiring and rather envious gentlemen.' According to *The*

World, she did not display 'a wince of fear or trepidation'. But we know from her own accounts that Nellie felt lost, dizzy and alone as the SS *Augusta Victoria* set out from New York Harbour.

That afternoon *The Evening World* carried a report of the liner's 9.40 am departure: 'The three stacks of the *Augusta Victoria* emitted puffs of black smoke and the great ship steamed majestically out into the North [now known as Hudson] River and passed down the bay and out through the narrows into the ocean.' The paper also printed praise and a prediction: 'Nellie Bly is probably the first of her sex to undertake it alone and unprotected, and it is no hazard to predict that her tour of the world for *The World* will become the most famous in the annals of travel.'

* * *

Alice and I have now tracked Nellie's departure from Hoboken on 14 November 1889; in our minds she is on her way to Southampton. The time has come to re-enact her arrival by rail in Jersey City, two miles downriver, on 25 January 1890. From Hoboken we take an eleven-minute train ride to Exchange Place in Jersey City, the spot where Nellie's railway car arrived exactly seventy-two days, six hours, eleven minutes and fourteen seconds after her journey began in Hoboken. Her train pulled into one of the world's busiest railway terminals for most of the nineteenth century. The once illustrious terminus where Nellie crossed the finish line served the ever-expanding Pennsylvania Railroad, as well ferries to Manhattan. Today, except for a few lines of disused railroad tracks embedded in nearby streets, you would never know it was ever here.

Back then, the piercing whistles of steam locomotives and double-decker ferries reverberated across the terminus where industry, commerce and transport converged. Yesterday's booming epicentre, Exchange Place feels eerily silent today. Svelte, grey high-rises stand back from the waterfront on this urban patch nicknamed Wall Street West for the financial institutions that cluster where trains and ferries once reigned. Like the City of London on a Sunday, there is little sign of life. It is strictly business as analysts, accountants and admin staff labour behind towering walls of tinted glass. Everyone is tucked indoors; Alice and I are the only ones outdoors. The steady drizzle that has pestered us all day clears at last, leaving satin puddles on the concrete that carpets this hushed financial district.

Where stillness now rules, pandemonium once prevailed. On Saturday 25 January 1890, when Nellie Bly's locomotive pulled up, 10,000 people swarmed the station to catch a glimpse of the triumphant globetrotter. 'It was a great human tidal wave', according to *The World*. 'When the train began slowly to enter the long, arched depot the assembled multitudes sent up cheer after cheer, and when the lithe little traveler stepped lightly from the train it was into the very arms of the surging crowd', *The World* reported. 'They cheered, they yelled, they hurrahed, they threw up their hats and fluttered handkerchiefs' for the sun-browned traveller with sparkling eyes. As soon as Nellie alighted, the three official timekeepers snapped their synchronised stopwatches. It was 3.51 pm. The journey round the earth was finished, *The World* reported. 'Phileas Fogg, the brave, the rich, the unconquerable, had been beaten – and by a woman.'

Nellie's ears rang with the roar of the crowds as she gazed across a sea of top hats, bowlers and feather-trimmed bonnets that flooded the vast station. Thousands were shouting themselves hoarse for the American girl who had turned an impossible dream into a reality. From the outside she seemed calm in her now-famous travelling costume, but inside Nellie was just as ecstatic as her adoring fans. 'I took off my cap and wanted to yell with the crowd, not because I had gone around the world in seventy-two days, but because I was home again,' she said. 'From Jersey to Jersey is around the world, and I am in Jersey now,' she cried out to her admirers.

No one had ever gone around the globe so fast. Across its front pages, *The Evening World* applauded Nellie. 'The completion this afternoon of Nellie Bly's wonderful journey ... will add one of the brightest pages to the imperishable record of the Achievements of Woman.' In a continuous tribute, the newspaper hailed her achievement: 'Without guide or escort; speaking no language but her mother tongue ... with but a single gown and an outfit which the ordinary woman would consider inadequate for a one day's visit to Newark, this frail, slender, plucky young woman has travelled over twenty-three thousand miles, has touched at every continent, has obtained flying glimpses of every phase of the world's civilizations, has demonstrated the perfection and simplicity of modern methods of travel, and has established a record which within her own lifetime would have been regarded as chimerical as a journey to the mountains of the moon.'

The official ceremony was led by the Mayor of Jersey City, Orestes Cleveland, who greeted Nellie with an enormous basket of flowers and an eloquent, if lengthy, speech. Cleveland, a manufacturer of black lead, stove polish and pencils, was one of New Jersey's leading politicians. He understood the importance of pomp and ceremony. Struggling to be heard over the gathered masses, the mayor congratulated Nellie on what she had achieved not only for females, but for all of humankind. 'The American girl can no longer be misunderstood. She will be recognised as pushing, determined, independent, able to take care of herself wherever she may go,' he said. 'People the world over have been taught that they are not so far apart as they had imagined and that is a great lesson.' Addressing Nellie directly, he said: 'You have set the whole world to thinking about it, and so have brought mankind nearer together.' The commotion was escalating, but the mayor carried on: 'You have added another spark to the great beacon light of American liberty that is leading the people of other nations in the grand march of civilization and progress,' he shouted. 'Passing rapidly by them, you have cried out in a language that all could understand, "Forward!" and you have made it the watchword of 1890.' The cheering crowds had now taken control, and before long Mayor Cleveland was forced to abandon his speech. With a police escort, he and Nellie snaked through the joyful mob to a carriage awaiting them at the ferry terminal. It was a Landau, a luxury carriage favoured by British royalty, especially for weddings, and Lord Mayors. Once it was driven down the gangplank and onto the boat, the Landau's folding top was thrown open so Nellie could stand, wave and bow to the 1,000 people squeezed on board. She did, all the way across the one-mile river stretch to the Cortlandt Street ferry depot in lower Manhattan. The landing there was brimming with exuberant crowds eager to greet the young circumnavigator. Her coachman was forced to drive the carriage 'like a plough' through the throngs, *The World* reported. The Landau proceeded up a seething Cortlandt Street and along Broadway, where the buildings were packed with people at windows hoping for a glimpse of Nellie Bly. Floral tributes, telegrams and colleagues awaited her arrival at

The World's offices around 4.30 pm for a reception in the main editorial room. A celebratory dinner was held afterwards at neighbouring Astor House, the nation's top hotel at the time.

The World's globe-girdler had accomplished what was 'incomparably the most remarkable of all feats of circumnavigation ever performed by a human being,' the newspaper declared. Nellie Bly is the 'best known and most widely talked of woman on earth today'.

Newspapers rejoiced the next morning: 'She's Broken Every Record! Father Time Outdone! The Whole Country Aglow with Intense Enthusiasm', blazed across the front page of *The New York World* on 26 January 1890. Alongside the jumbo headlines, a cartoon running across four columns placed Nellie at the head of history's great navigators, from Sir Francis Drake to Captain Cook and of course the inestimable Phileas Fogg. 'A Little Pardonable Consternation Among the Globe-Circlers at the Remarkable Achievement of "The World's" Traveller,' read the caption. Below the cartoon, a special cable dispatch report from London correspondent Tracy Greaves announced 'All Europe Enthusiastic. Congratulations from Geographers, Scientists and Friends'. Nellie's success was now known across the length and breadth of Europe. 'Here in England the trip excited an unusual amount of interest perhaps because at the outset the papers had predicted Miss Bly would not succeed,' Greaves wrote. 'But they paragraphed her various stages along the journey and all give her credit now for her pluck and perseverance.'

In Paris and Amiens, correspondent Robert H. Sherard had gathered commendations for Nellie and *The World* from no less than French President Sadi Carnot, the Vicomte de Lesseps, developer of the Suez Canal, and the Vicomtesse; as well as Jules and Honorine Verne, who said they were ecstatic.

Telegrams flew into the London Bureau of *The New York World*; three were published in the newspaper. A message from Jules Verne, printed in French and English, said: 'I never doubted the success of Nellie Bly. She has proved her intrepidity and courage. Hurrah for her and the Director of THE WORLD! Hurrah! Hurrah!!' From P&O's London manager F. H. Fifth: 'There are comparatively few people who will read of it who will know the real difficulties and hardships of her journey or will thoroughly comprehend what a tremendous distance she has travelled in so short a time.' Even P.T. Barnum of Barnum and Bailey Circus had something to say: 'As an American who knows by experience the difficulties and dangers of travel, I congratulate Miss Bly, of THE WORLD, on her most remarkable success, personal pluck and courage,' the 79-year-old showman wrote. 'From the great amount of public attention she is attracting both at home and abroad I begin to look in her direction for more popular features in my show.'

For a statement from the president of the Royal Geographical Society (RGS), Tracy Greaves called on the splendidly named Sir Mountstuart Elphinstone Grant Duff at York House, his seventeenth-century stately home in Twickenham, now the Town Hall of the London Borough of Richmond upon Thames. 'For my part, I think it best in travelling to see foreign countries slowly, but if any more enterprising Americans desire to emulate Miss Bly's example, it is much better to travel rapidly than not to travel at all,' the white-bearded author, politician and former Governor of Madras told Greaves. 'Miss Bly has proved herself a remarkable woman and I hope she will get a good husband.'

Despite his somewhat patronising praise, implying that circling the world was merely a means of attracting a husband, it turns out that Grant Duff was actually a champion for women, at least women explorers and geographers. He was the first to open the RGS doors to female Fellows. His proposal to grant women the same rights as men was carried almost unanimously in 1892, the society archives reveal. When it was revoked amid controversy a year later, the RGS President resigned from his prestigious post, leaving another twenty years before women could be admitted as Fellows. More than a century later, thanks to my RGS-listed Nellie Bly expedition, I would become a Fellow. In 2018, the first-ever conference hosted at the society on 'The Heritage of Women in Exploration' honoured female adventurers past and present including Marianne North (1830–90) whose 832 botanical paintings from exotic travels hang at Kew Gardens; Alexandra David-Néel (1868–1969), said to be the first Western woman to enter the holy city of Lhasa; today's celebrated polar explorer Felicity Aston; and of course Nellie Bly. The conference was arranged by Women's Adventure Expo, a social enterprise adventure hub for women, of which I am a founding trustee.

* * *

Alice and I have one more journey to complete before we can wrap up our re-enactment of the first and last days of Nellie's record-breaking voyage. On board the 4.17 pm New York Waterways commuter ferry from Jersey City, we are tracing her jubilant victory parade across the Hudson River to Manhattan. Nellie shared her ferry with a thousand well-wishers; we are travelling with eight other passengers in this pre-rush hour slot. The skyline views, obscured by the rain when we crossed the river this morning, now glisten on both sides of the Hudson. Behind us, bands of billowing wake trail the ferry like banners as the waterfronts of Jersey City and Exchange Place fall away. Ahead, the ultimate panorama of Manhattan unfurls. New York City's tallest building, the eight-sided One World Trade Center in Lower Manhattan, whirls skyward in a tower of isosceles triangles. To the left in Midtown Manhattan, Art Deco masterpieces like the Empire State Building, the world's most-loved skyscraper, and the Chrysler Building, its prettiest, command the view.

When Nellie crossed the Hudson that historic day, she was too busy waving to notice the skyline. Even if she had, her view would have been void of towering pinnacles. It was her boss Joseph Pulitzer who sparked the skyscraper phenomenon in New York City. When the twenty-storey World Building, headquarters for his newspaper, was completed in December 1890, it was the tallest in the city, and indeed the world. It was under construction when Nellie returned from her global adventure.

The river crossing behind us now, Alice and I land back at Pier Eleven where we began our mission this morning. Now on foot, we are following Nellie's carriage route along Cortlandt Street up to Broadway. Traffic, not people, chokes the streets as we try to imagine Nellie's Landau plying the crowds and dodging the horse-drawn streetcars halted in their tracks by the day's festivities.

In eight hours, Alice and I have attempted to re-create the two most momentous days of Nellie's fabled journey. For me it is an anti-climax. I cannot place Nellie. She is not here. Like the steam-powered ocean liners on her journey, most tangible traces of Nellie's presence have disappeared. Yesterday's plucky globetrotter has been displaced by today's reality. There are exceptions. I could 'see' her in London's Charing Cross Hotel as she grabbed a quick cup of coffee before boarding the train to Folkestone. She lives on in Jules Verne's home in Amiens where I joined her in his study. We slept in the same hotel – the Grand Oriental – in her Ceylon and my Sri Lanka. In Singapore we both paid our respects to the statue of Sir Stamford Raffles. We rode the Peak Tram in Hong Kong, and stepped inside the Great Buddha in Kamakura, Japan. But time has robbed us of many sites standing just 125 years ago, especially here in New York City. The grand train stations and ferry depots of her day were demolished decades ago. Even the famous *New York World* building is gone. It was razed in 1955 to expand entry ramps onto the Brooklyn Bridge.

Still, one building readily associated with Nellie Bly remains in Manhattan after decades of neglect. It is the scene of the first story she ever wrote for *The New York World*. Today it is known as the Octagon Tower, now part of a high-end residential community that invites you to 'experience this sanctuary' on Roosevelt Island. In Nellie's day it was a 'living hell' on what was then called Blackwell's Island, home of the New York City Lunatic Asylum. It was here that Nellie Bly went undercover to expose the barbaric treatment suffered by female patients.

Nellie was not the first to notice. On his tour of America in 1842, Charles Dickens visited the asylum and found that 'everything had a lounging, listless, madhouse air, which was very painful. The moping idiot, cowering down with long disheveled hair; the gibbering maniac, with his hideous laugh and pointed finger; the vacant eye, the fierce wild face, the gloomy picking of the hands and lips, and munching of the nails: there they were all, without disguise, in naked ugliness and horror.' Although alarmed by the abysmal conditions at the asylum, Dickens admired its handsome architecture and noted its 'spacious and elegant staircase'.[1]

Dickens was an official guest at the asylum. Forty-five years later, Nellie entered as an inmate after convincing authorities that she was mad. She arrived on the island in a filthy boat reeking of sweat, tobacco and urine to enter the asylum in September 1887. 'What is this place?' she asked the man dragging her off the boat. 'Blackwells Island, an insane place, where you'll never get out of,' he replied.[2] Ten days later she was out. Nellie's asylum exposés rocked the nation, and Blackwell's Island became the birthplace of investigative reporting. That is why Alice and I decide to visit this canoe-shaped island anchored in the East River between

Manhattan's Upper East Side and Queens. It is not part of my official round-the-world trail, but we must go where Nellie made her name as a journalist.

We travel over on a cherry-red Roosevelt Island Aerial Tram that might look more at home in a ski resort. On this Friday afternoon, we gaze down on a city preparing for the weekend as we sail 250 feet high in the gondola-style cable car that will drop us in the middle of the two-mile long island. From there we will follow Main Street to the northern end of the island to reach the site of the former lunatic asylum. A five-storey, eight-sided rotunda built of metamorphic stone quarried on the island is all that is left of the multi-winged asylum that once housed 1,300 desperate patients. This was the main entrance to Nellie's madhouse. Innocent souls left their freedom, and often their sanity, here on this doorstep for more than half a century. You would never know by looking at this impeccably restored rotunda, the newly-named Octagon Tower, serving as the centrepiece of an exclusive 500-unit apartment complex occupied by young professionals and retirees. The street number 888 is chiselled in stone above a recessed doorway embraced by a flying double staircase. Capped with a hexagonal fish-scale dome, the rotunda dates back to 1834. To the left of the entrance, raised gold letters on a cast bronze plaque spell out an abbreviated history of today's Octagon Tower, and announce that it is a city, state and national landmark.

I am slightly apprehensive that anguished spirits may linger inside, but Alice and I enter the rotunda turned tower, because Nellie did. The ghosts are gone, I think. A sweeping spiral staircase engulfs the space, orbiting gracefully up to the dome. Today's streamlined version of winding metal and polished timber has replaced the decrepit wooden stairs, rails and balustrades that Nellie would have known.

When we were here there was no mention of Nellie in this former lunatic asylum, or anywhere on the island as far as we could gather. Not anymore. The island has embraced her. More than 130 years after her break-through exposés of the asylum, Nellie Bly is returning to the island in a memorial that will celebrate her legacy as a journalist and a humanitarian. The installation, designed by Amanda Matthews of Prometheus Art in Lexington, Kentucky, is entitled *The Girl Puzzle*, named for Nellie's first newspaper article, the story that launched her journalism career, in which she pleaded for opportunities for girls and women. It was published in *The Pittsburg Dispatch* on 25 January 1885. Four years later, on exactly the same date, Nellie captured global headlines when she won her race around the world. The journalist's life and legacy will unroll along a sixty-foot walkway in Lighthouse Park, not far from the asylum that she unmasked. Four monumental bronze female faces, representing the vulnerable women Nellie championed, will be overseen by the crusading journalist herself, her face cast in silver bronze. Three mirrored spheres will portray the significant phases of Nellie's life: her early career in journalism, her impersonation of madness to gain entry into the asylum, and her seventy-two day trip around the world.

Beneath Nellie's silvered face, the artist plans to include a tribute to her from Arthur Brisbane, one of the best-known American newspaper editors of the twentieth century.

'Nellie Bly was THE BEST REPORTER IN AMERICA and that is saying a good deal … She takes with her from this earth all that she cared for, an honorable name, the respect and affection of her fellow workers, the memory of good fights well fought and of many good deeds never to be forgotten by those that had no friend but Nellie Bly. Happy the man or woman that can leave as good a record,' Brisbane wrote in the *New York Evening Journal* on the day after her death.

'Nellie Bly told the stories of other women,' says Amanda Matthews of the installation that will honour Nellie. 'Now, we will tell hers.'

The Girl Puzzle commemorates Nellie Bly's legacy to journalism. Shelton J. Haynes of the Roosevelt Island Operating Corporation, commissioners of the installation, says: 'The significant role that journalism has played in calling to attention societal ills has always been etched in the fabric of change. Nellie Bly, in particular, was a stalwart investigative journalist who began her impressive career on Roosevelt Island. Her journalistic contributions make her a pioneer and serve as a blueprint for all journalists.'

Chapter 13

In Which Nellie is Laid to Rest

DEDICATED JUNE 22,1978
TO
NELLIE BLY
ELIZABETH COCHRANE SEAMAN
BY THE NEW YORK PRESS CLUB
IN HONOR OF
A FAMOUS NEWS REPORTER
MAY 5, 1864 ~ JAN 27,1922.

Nellie Bly's headstone, Woodlawn Cemetery

New York City
27 January 1922

'Nellie Bly was THE BEST REPORTER IN AMERICA ...'
Arthur Brisbane, editor, *New York Evening Journal*

Woodlawn Cemetery is forty minutes by car from Midtown Manhattan along New York State Road 9A, heading north along the Hudson River to the Bronx. Sally is driving us there. Our Metro cards for the subway, buses and even the aerial tram to Roosevelt Island remain in our pockets. They have served us well in our quest to track Nellie in New York City. Alas, there are no more footsteps to follow. Nellie has been buried in Woodlawn Cemetery for almost a century. After so much time on public transport, it is a treat to leave the city by car on a fourteen-mile excursion, a field trip into the 'country'.

The Bronx was the countryside in 1863 when Woodlawn Cemetery was being established. The farmland site was chosen by a group of big-name New Yorkers who envisaged a rural resting place with plenty of room for visitors to move between mausoleums, monuments and headstones. Subsumed by the encroaching metropolis over the decades, the Bronx today is New York City's third most crowded borough, the home of

Yankee Stadium and the birthplace of hip-hop. Amid this urban clamour, Woodlawn Cemetery and Conservancy remains a 400-acre oasis of calm where life and death go hand in hand.

Sally, Alice and I pull in and head straight for the Gothic-style cemetery headquarters where Susan Olsen, doyenne of Woodlawn's history, is expecting us. Sally made our appointment months ago. Emerging from her office with a notebook and a handful of files, Susan has mapped out a trail across the city's largest cemetery to find Nellie's grave in the south-eastern section. We are just three among the 100,000 international visitors to Woodlawn every year. Even so, we get the VIP treatment from Susan. Black designer sunglasses are propped amid her honey-blonde hair to keep it off her forehead. Her cemetery 'uniform' consists of a navy Woodlawn-branded polo shirt, khaki trousers and sneakers for traversing the tracks and looping roads we will follow. Keeper of the cemetery's heritage for more than thirteen years when we meet, she seems just as excited as we are to visit Nellie's final resting place, yet again. Susan is part of the landscape here, just as rooted as the woodland trees that shade our path.

The late September sun casts shadows on the tombstones surrounding us as we navigate this multi-faith cemetery that feels more like an open-air museum. More than 300,000 bodies rest in peace here. Alongside the lavish memorials and simple headstones, Woodlawn is a wildlife refuge with woodpeckers, doves, geese, chipmunks, even a lake. It is also an official arboretum with 140 varieties of trees. Today some species – the beeches, oaks, maples and sassafras – are speckled with ruby and gold, hints of the radiant autumn to come.

The National Park Service describes the cemetery as 'a popular final resting place for the famous and powerful'. Luminaries from the worlds of music, literature, politics and industry are buried here, from Gilded Age magnates such as F. W. Woolworth to jazz greats such as Duke Ellington, and anti-slavery activists such as Elizabeth Cady Stanton.

Like a cast from a movie, key characters in the story of Nellie Bly's world journey rest with her at Woodlawn. The father of American journalism, founder of the Pulitzer Prize and Nellie's former boss, Joseph Pulitzer, is commemorated in a family monument featuring a life-size bronze of a mythical male in classical garb, seated in contemplation. There are no words. Embedded in the lawn in front of the monument, four stone rectangles mark the Pulitzer family graves. An 11-inch American flag, planted at an angle in the grass, flies over the newspaper tycoon's stone, engraved simply with JOSEPH PULITZER, BORN APR 10, 1847; DIED OCT 29, 1911. Not far away, Nellie's fellow world circumnavigator Elizabeth Bisland lies with her husband Charles Whitman Wetmore under a triptych headstone decorated with stylised vines, lilies and a garlanded urn. The vines symbolise friendship, the lilies represent purity, and the urn immortality. We pose for photographs around her tomb, feeling the warmth released from the sun-baked granite. I place one of twelve long-stemmed white roses – from a bouquet I brought for Nellie – on Elizabeth's grave. Few remember that Elizabeth Bisland also beat Phileas Fogg's eighty-day record, finishing her global journey in seventy-six days. As a self-proclaimed 'lady of literature' who avoided the headlines, I do not think Elizabeth would mind. Although they shared an astonishing world adventure, the two journalists never connected.

With leafy paths winding amid castle-like mausoleums, tree-shaded headstones and larger-than-life marble angels, Woodlawn would surely remind Nellie of the Happy Valley cemetery that she so admired in Hong Kong. They are both archetypal garden cemeteries set in acres of landscaped countryside. Woodlawn claims to be one of America's finest. Visiting Hong Kong's Happy Valley in December 1889, Nellie felt she was in a public garden, even a park. By the time I got there 125 years later, the burial grounds were derelict and crumbling. Even though the two cemeteries are continents apart, I welcome this chance to understand what fascinated Nellie about Happy Valley. Based on her enthusiasm for the cemetery there, I like to think that she would find Woodlawn a suitable place to be buried.

At last we are nearing Nellie's burial site – grave 212, section 19 in the Honeysuckle Lot. As we approach, Susan sets the scene of Nellie's burial. She has retrieved the official paperwork to show us. Passing the records to me, Woodlawn's historian explains that Nellie's family and friends did not provide sufficient funds for her interment. It seems Nellie herself did not set any money aside. She had none, we now learn. I am not prepared to find Nellie buried in a pauper's grave. As the reality of her final days becomes clear, I begin to understand that this 'pilgrimage' will not be the culmination of my journey with Nellie that I had expected. My eyes sting, uninvited tears well and fall, and my lips tremble, not so much with sadness, but with a bitter sense of injustice for Nellie. She died of pneumonia on 27 January 1922, aged 57, at St Mark's, a hospital in Manhattan founded to treat impoverished people. Two days later, Nellie was buried in a simple grave among the victims of the Great Influenza Epidemic of 1918–1919, the deadliest in history. She laid in virtual obscurity for fifty-six years until the New York Press Club erected a modest granite headstone.

DEDICATED JUNE 22, 1978
TO
NELLIE BLY
ELIZABETH COCHRANE SEAMAN
BY THE NEW YORK PRESS CLUB
IN HONOR OF
A FAMOUS NEWS REPORTER
MAY 5, 1864 – JANUARY 27, 1922

Nellie is much more, however, than a famous news reporter. She paved the way for female journalists, opening newsroom doors to women like me, and pioneered investigative reporting – the kind of journalism that brings about change and makes the world she circled a better place. Nellie Bly defied the status quo, gave voices to the vulnerable, championed women's rights and challenges us, still today, to believe that 'nothing is impossible if one applies a certain amount of energy in the right direction'.

We stand before her simple headstone, no higher than our knees, to pose for photographs. I arrange the eleven remaining long-stemmed white roses on the top of her rounded headstone. Before we leave, I will place the roses at the base so they do not blow away. Their fragrance is soothing. I am trying to be discreet but the tears flow freely. I use the front of my hand to wipe them away. It is a silent expression of unexpected grief. Alice knows I am struggling. Maybe Sally does too. A few tawny maple leaves float over on a soft wind and settle in the grass around us. We have paid our respects to Nellie.

Chapter 14

In Which Nellie is Born to Be a Journalist

Nellie Bly's birthplace, Cochran's Mills

Cochran's Mills and Apollo

'A crusading journalist on Pittsburgh and New York newspapers, she won fame for her daring exploits and her investigations of social ills. In 1889–90, Bly circled the world in 72 days. She was born Elizabeth Cochran and lived here as a child'.

Pennsylvania State Historical Marker

The people of Apollo, Pennsylvania never forget Nellie Bly's birthday. Every year on the Sunday closest to 5 May, the local historical society hosts a party for the whole town complete with a homemade cake, candles and entertainment such as one-act skits and talks by loosely related descendants. Not a year has gone by without a party, at least in the last two decades. Her 150th birthday in 2014 was celebrated with a multi-media show and quiz presided over by Apollo-Ridge High School drama club students portraying Nellie and fellow globetrotter Phileas Fogg. An account of the festivities made it into *The Valley News Dispatch* and out into cyberspace where I seized it – and decided to add Nellie Bly's hometown to my itinerary.

By now my 'official' world journey is complete. I am free to detour off the beaten track – to travel from where Nellie's life ended to where it began. My husband David has flown across the Atlantic to join me for a cross-country road trip to visit her childhood home in Apollo and her birthplace in Cochran's Mills, both bounded by the hills and rivers of Armstrong County.

Interstate 80, an East-West freeway spanning the USA, leads most of the way between Nellie Bly's last home in New York City and her first homes in southwest Pennsylvania – a 350-mile drive that skirts the fifty-seven years of her life. It just so happens that the 'pilgrimage' I have tacked onto my travels will coincide with the Apollo Area Historical Society (AAHS) Sunday Evening Programme Series on 6 October. I notice it on the society's Facebook page and get in touch with AAHS vice president Sue Ott. She immediately invites me to speak, and mentions that we would be eligible for the historical society discount at Dolly's Guest House if we want to stay the weekend. We do. A long-serving AAHS board member, Dolly Lackey McCoy, runs the country-style bed and breakfast on River Road in North Apollo.

*   *   *

We pull into Dolly's Guest House, which is dozing under shade trees at the back of a vast front yard. A three-seat swing chair sits on the lawn. We follow a curved concrete path past a Victorian lamppost to a canopied front porch set with a pair of white iron filigreed chairs. A Pennsylvania black bear, sculpted from a tree trunk with a chain saw, greets us as we approach. He holds a sign in his hefty paws with 'Welcome' branded into the wood.

Dolly hears our footsteps and opens the front door to usher us in. She knows exactly why we are here, and she is at our service. Our hostess is wearing a burgundy jersey embroidered with autumn leaves and miniature pumpkins. They like to celebrate the seasons here, especially fall. Her substantial strawberry blonde hair is gathered at the back and bundled into a beige scrunchie. A spongy fringe frames a broad face sprinkled with fading freckles. Dolly's eyes almost get lost whenever she smiles, which is often. With spindle bedsteads, rag dolls, crocheted doilies and dishes of peanut butter candy, staying at Dolly's is like visiting your grandma's house. The beds are spread with handmade patchwork quilts collected by this Apollo native who loves antiques and the sort of knick-knacks that make a place feel cosy – piggy banks, ceramic dolls and vases bursting with artificial flowers. We deposit our bags in the bedroom and return to Dolly's kitchen for a briefing. Here at Dolly's Guest House, we are eight miles from Nellie's birthplace; but nothing is left of it, she tells us. Cochran's Mills in Burrell Township has disappeared – been swept away by man-made floods, swallowed by hallowed pinewoods, totally lost in time. A few obstinate foundation stones are the only vestiges of this former mill community, and they take some finding. There is only one person in the whole of Armstrong County, even the entire state of Pennsylvania, who can trace the ghostly footprint of Cochran's Mills, according to Dolly, and that is Arnold Blystone, co-founder of the Burrell Township Historical Society.

Before we have finished the pre-sweetened iced tea that Dolly poured for us, she is on the phone to Arnold Blystone. Sure enough, he will be happy to meet us after breakfast the next morning at the Burrell Township Municipal Building and Volunteer Fire Department, which doubles as headquarters for his historical society. Dolly is coming along.

It is shortly after 8.00 am and Arnold is waiting for us in his oversized pick-up truck as we drive into the parking lot. He is a stocky man, just the right weight to fill his looming frame, with watery blue eyes like Dolly's, a face ruddy from a lifetime outdoors and an easy smile that makes you like him straight away. Tufts of white hair escape from a camouflage cap advertising Buffalo Limestone Inc. Arnold wears his almanac knowledge lightly, and he shares it readily.

Burrell Townships' purpose-built municipal building is larger than a barn, with a pre-fab feel to it. One maize-coloured wall is devoted to monochrome photographs, all neatly mounted in mismatched frames with labels cranked out of someone's word processor. Horses and buggies, family portraits, churches and snow scenes reside in tidy rows across the wall; and most importantly for me, the lost settlement of Cochran's Mills. There it is, in blurry black and white. A handful of clapboard houses straddle a wandering stream. The tallest of the buildings, what must surely be the grist mill, hugs the left bank just after the wooden bridge where Crooked Creek turns back on itself. No more now. Nothing.

Following our indoor orientation, Arnold Blystone leads us outdoors, deep into the musky woods that claimed Nellie's birthplace. Striding past Crooked Creek and straight through two centuries, Arnold knows exactly where the remaining foundation stones lie. With his steel toe lineman boots, he kicks away layers of gunky brown sludge to reveal a set of slab-like stones strangled in black moss. All the while he is painting word pictures of Cochran's Mills, back when Crooked Creek flowed freely. Arnold knows where the biggest bass were caught, where the blacksmiths built their forges, where the local doctor lived and where the three mills once toiled.

Except for patches cleared for homesteads, broadleaf forests of maple, cherry, beech and oak tumbled down from the hills right up to the banks of Crooked Creek, where Cochran's Mills perched on one of its biggest bends. Fields of corn, rye, wheat and oats spread like blankets around outlying farms, their harvests destined for the local mills. Earthen tracks led between the houses, crossing to the general store and post office, and over to the creek-powered mills – a four-storey grist mill for flour and corn meal, a sawmill, and a fulling mill for cleansing and thickening cloth. Nurtured by their creek, their forests and their farms, the folks of Cochran's Mills ground out a living in some of the oldest human occupations – millers, farmers, stonemasons, and later blacksmiths, merchants, wagon-makers. In the midst of the spinning blades and rasping millstones stood Nellie's father Michael Cochran, leader of the community that came to bear his name. The mills prospered and so did Nellie's father, later serving as local postmaster, and associate justice of Armstrong County. They called him Judge Cochran, and still do.

Nellie was born Elizabeth Jane Cochran here on 5 May 1864. It was the beginning of the end of the American Civil War. Before she was a year old, the Confederates would surrender, ending the bloodiest four years in US history, and Abraham Lincoln would become the first American president to be assassinated. Her first five years were spent in a two-storey, unpainted frame house where Cherry Run flowed into Crooked Creek. It had a gabled roof with two brick chimneys, one at each end; lattice-fenced front verandas ran along both floors. Six wide wooden steps led up to the entrance. Family and friends used them as seats to take the weight off their feet, and sit and chat awhile. Towering oaks and sycamores shaded the house on three sides; a clothesline was strung across the front. On the west side, an axe was lodged into a chopping block to cut wood for the fire and the stove.

Cochran's Mills sustained Nellie's mother and the fifteen children Michael fathered. Nellie was the thirteenth child, and the third born to Michael's second wife Mary Jane, whom he married in 1858 after he was widowed. Nellie had one sister and three brothers, five half-sisters and five half-brothers. I had but one sister, an artist, Pauline, who we lost to multiple sclerosis in 1982 when she was 26. This book is for her.

Even with our eyes squeezed tight, in the deep stillness of these backwoods, it is hard to conjure up scenes of corn being crushed under massive stones, sawdust spraying from squealing blades, and creek waters churning fibres into felt. Long gone are the days when the air hung heavy with the sweet scents of sap, newly hewn boards, and freshly milled cornmeal. The millstones turn no more, but there is one that remains and Arnold wants to show it to us. The mid-morning sunshine illuminates the woodlands we cross on foot, offering a sneak preview of the autumn colours that will soon unfold. Just ahead of us, erect and solitary like a Neolithic monolith, a nineteenth-century millstone is anchored in the earth. Tickled by geraniums and cornflowers, it stands in the middle of a freshly mown lawn scattered with grass clippings that smell like the colour of green. A bronze plaque is embedded in the millstone's centre where grooves scored for the grain fan out like sunbursts. In raised gold letters it reads 'In Honor of Nellie Bly' and sets out her achievements as a journalist, humanitarian and world traveller. They are cast in perpetuity thanks to the Nellie Bly Questors Chapter #533 of Armstrong County, Pennsylvania. A global organisation, the Questors aim to 'keep history alive' through preservation, restoration and education. The local chapter's loyalty to Nellie is evident in the 1867 birth date they inscribed on her memorial, extending her youth by three years. Nellie would love it. It is the same birth date she swore to when authenticating her round-the-world passport in London, when she chose to be 22 instead of 25. Arnold Blystone, Dolly and I take turns posing for photos at the 'Bly stone', resting our forearms on the cool, honed limestone now serving a commemorative role since its retirement from the mill. David is taking the pictures. Now that we have seen the traces of Nellie's birthplace, it becomes time to continue our journey through her childhood. We say our farewells to Arnold and head east through Armstrong County to the town of Apollo, just as Nellie and her family did 145 years earlier.

The Judge was coming home. The son of Irish settlers, he was raised in Apollo from the age of 4. Nellie was 5 when her family took up residence in the impressive two-and-a-half storey Greek revival mansion that he built for them on Terrace Avenue in the north of town. It stands to this day. The towering white portico for carriages that rose from the front of the house has disappeared, along with all of the outbuildings on the original three-acre plot. Even so, Nellie would recognise her second home, now sitting dejected on a fraction of its land. Despite blistering white paint, missing shingles and patches of rotten wood, its former grandeur shines through. A pedimented ionic porch, festooned in carved garlands and bows, leads onto a column-fronted veranda enclosed by a grey balustrade. Matching triptych windows of iridescent glass flank the front entrance. An American eagle spreads its brass wings above the panelled front door and below the house number, 505. To the side, four black rusting letter boxes labelled A, B, C and D are nailed to the wall for the apartment dwellers that now occupy Nellie's home. Although decidedly worn, number 505 takes its place in a parade of historic houses on Terrace Avenue, once known as Mansion Row, a testimony to Apollo's one-time age of prosperity.

But crushing poverty, not prosperity, was on the cards for Nellie and her family. She was only 6 when her beloved father died without warning, catapulting his once wealthy and respected family into unforeseen hardship. He did not leave a will. In less than a year, the mansion the Judge had provided for his new family was sold out from underneath them. His widow's attempt to safeguard her children by re-marrying ended in domestic brutality and divorce. At 14, Nellie found herself in the County Court recounting the violence inflicted on her mother by her stepfather. When she was 16, the shame imposed by death, cruelty and divorce drove the family out of Apollo and into the grimy, soot-choked suburbs of Pittsburgh. What Nellie lost in security and standing with the death of her father, she seems to have gained in sheer resolve, enterprise and courage. The indignity borne by her family forged a determination to triumph over tragedy and fight for justice, especially for the most vulnerable. It became her burning mission and she employed it throughout her life. With that searing sense of injustice, Nellie thrust open the doors to journalism.

Staked in the grass facing the 'mansion' where Nellie survived one of the most cataclysmic events of her life, a state historical marker honours her legacy. 'A crusading journalist on Pittsburgh and New York newspapers, she won fame for her daring exploits and her investigations of social ills. In 1889–90, Bly circled the world in 72 days. She was born Elizabeth Cochran and lived here as a child.' It is the essence of Nellie Bly.

As the historical markers here in Apollo and at Cochran's Mills attest, Nellie had a thing about names. She had four. Born Elizabeth Jane Cochran, as a child she was called Pink, sometimes Pinky, after the rosy outfits her mother chose for her instead of the dingy browns and greys worn by other little girls. In 1879, at the age of 15, an aspiring Nellie added an e to the end of her last name to make it more distinctive. Her *nom de plume*, chosen by Nellie's managing editor at *The Pittsburg*[1] *Dispatch*, was borrowed from a minstrel song by American songwriter Stephen Foster.[2] Female journalists in those days could not write under their legal names. When Nellie married millionaire industrialist John Seaman in 1895, she became Elizabeth Cochrane Seaman. By any name, these two memorials stand as tangible evidence of the pride Pennsylvanians take in 'their Nellie Bly'.

Apollo was Nellie's home for a decade. To this day, her childhood and subsequent lifetime achievements are one of the town's main claims to fame. Thirty-five miles northeast of Pittsburgh, it sits on the north bank of the Kiskiminetas River, a dozen miles from where it becomes a major tributary of the mighty Allegheny River. Folks around here merged the first two syllables of each river and declared their region the Alle-Kiski Valley (AKV). There is the AKV News, the AKV Historical Society, even an AKV Chamber of Commerce. Nellie is an 'AKV girl'. These two rivers define the lives of the valley dwellers. They are constant, unlike the coal and steel industries that once surged across the region. It is hard to fathom that together with Pittsburgh, the Alle-Kiski Valley mill towns, including Apollo, ranked as the steel-making capital of the world for more than a century, from 1875 to 1980. The Brooklyn Bridge, the Empire State Building, the lock gates of the Panama Canal and the axle for the world's first Ferris wheel are among the iconic structures forged from their steel. Four decades have passed since blast furnaces belched fire and smoke across the skies of these mill towns, when the steel belt, defeated by globalisation, collapsed into the rust belt. Wasted coalfields, abandoned mills and lifeless town centres smoulder in the shadows of Pittsburgh as steel towns strive to reinvent themselves. Apollo took a big hit; but it does not feel like it, at least when we are here.

A gentle pride prevails, perhaps best captured by Dr T. J. Henry, physician, surgeon and a classmate of Nellie's,[3] in his 1916 *History of Apollo Pennsylvania, the Year of a Hundred Years*. He begins his centennial book by defending why he wrote it. 'Importance is relative. It is not necessary to be a city of the first class to fill the niche in the hearts of the people or the history of the state. Besides it is our town … our home.' Nobody seems quite certain who chose the name Apollo, but there is a pretty good chance it was resident physician, poet and student of classics, Dr Robert McKissen, whose 'knowledge of Latin and Greek was better than average', according to Henry's centennial history. Even more than its association with the celebrated Olympian god, townsfolk love their name partly because Apollo PA (abbreviation for Pennsylvania) forms a palindrome; words that read the same forwards as backwards like race car, civic and kayak. But what really makes them stand tall is its association with the planet's most astonishing space mission ever, the Apollo 11 lunar landing; and the fact that their county, Armstrong, is the last name of the first man to walk on the moon.

It did not take rocket science for town officials to grasp the potential. By the time Neil Armstrong had taken his 'one small step for man and one giant leap for mankind' on 20 July 1969, Apollo had embraced the moon landing as its very own. That same day, members of the volunteer fire department donned spacesuits and drove to Moon Township in the next county where they planted a flag and came back with 'moon soil'. The three astronauts, Armstrong, Buzz Aldrin and Michael Collins, became honorary citizens, and Apollo Astronaut's Way took its place on the street map. That same 'moon soil', packed into a three-legged pressed glass pot, is on display at the state's largest history museum, Pittsburgh's Heinz History Center. Apollo's historical museum boasts its own collection of moon memorabilia – a thank you note from Neil Armstrong on National Aeronautics and Space Administration letterhead, Man on the Moon newspapers, and collectable caps, plates, mugs and glasses bearing the astronaut's faces, the moonwalk, President Kennedy and the US flag. Mock spacesuits hang ready to be worn by fire-fighters for the annual Apollo Moon Landing celebrations that began the day after the Lunar Module Eagle landed in 1969, and, like Nellie's birthday celebrations, still continue, give or take a year or two.

Here in Apollo, population 1,550, it feels as if most people know each other or may even be related. Neighbours tend to gather at Lackey's Dairy Queen or the Yakkity Yak Diner. The Dairy Queen opened on 3 August 1955 with two-for-one ice cream sundaes and live South Sea Island music. It is a mainstay of the community; many of Apollo's youngsters worked at 'the home of the cone with a curl on top' to help pay for college or their first car. Our hostess Dolly, a Lackey before she married, grew up in a world of ever-flowing, soft swirl ice cream. Yesterday's neon-rich drive-in, crowned by a colossal vanilla cone, has been re-modelled with a barn-like façade, a shingle roof, and a red and white awning announcing 'Lackey's Dairy Queen Since 1955'. Dairy Queen is now DQ and the Ford Fairlanes, Studebaker Hawks and Chevy Bel Airs that once flocked here have been replaced by hatchbacks, pick-up trucks and SUVs. Autumn is just around the corner, and the DQ is preparing to shut down for the season. People here are stocking up on ice cream supplies for their freezers, ready for the harsh Pennsylvania winter ahead. They know it will be a zinger because so many furry black caterpillars have been spotted, an omen in these parts.

The Yak Diner nearby operates all-year, daily from 7.00 am to 8.00 pm, except on the first day of deer-hunting season when it opens at 4.00 am. Dolly sends David and me there for dinner before the start of the historical society meeting at 6.30 pm where I will recount my Nellie Bly adventure. Streamlined stainless steel inside and out, the Yak, a 1955 O'Mahoney dining car, is renowned for the Yak Mak, its version of the McDonald's favourite. The Yakkity Yak, a fried stacked bologna sandwich, is another favourite, along with the crispy chicken stuffing stack available 'only at the Yak', not to mention blue plate specials like meatloaf, liver and onions, and corned beef hash. Rather than teetering on swivel stools parked like toadstools along the lunch counter, or settling in a burgundy Naugahyde booth, we choose a Formica-topped table for two, with Early American chairs. It is not far from the jukebox. In honeyed vocals, Dolly Parton is begging Jolene not to steal her man. 'Please don't take him just because you can,' she pleads, as the 45 rpm record spins. David orders corned beef hash with coleslaw and apple sauce. I dismiss comfort food and go for a 'ladies who lunch' Cobb salad – iceberg lettuce, tomato, hard-boiled eggs and Roquefort cheese – accompanied by bottomless cups of coffee, a diner staple. Digging into my classic salad drenched in blue cheese dressing, I watch the time on the circular Art Deco stainless steel clock above the lunch counter. We need to allow a good ten minutes to drive from the Yak on River Road to the meeting in the former home of the Women's Christian Temperance Union (WCTU) on North Second Street, headquarters of the AAHS since 1998.

The milky-white two-storey clapboard building, across from the New Life Baptist Church, stands where it was built more than a century ago. The paint is so fresh you can almost smell it. The stars and stripes wave from a flagpole fixed to the temple-fronted façade. It is just the kind of place you would find in a Norman Rockwell print celebrating the Fourth of July or Memorial Day. Embellished with fluted ionic columns and a grand arched entrance, the WCTU building is said to be the best example of Greek revival architecture in town. The Apollo branch, organised by local matrons to combat the evils of alcohol, dates back to 1889, the same year that Nellie was circling the world. Ten years later, the building was erected under the direction of Mrs Mary J. Guthrie, WCTU president for twenty-four years.

David and I climb the five steps past the American flag outside and head straight upstairs to the first-floor meeting hall, where an audience of about forty is settling into a semi-circle of metal folding chairs. The aroma of coffee floats up, a pleasing reminder that refreshments will be served after the meeting. The hall darkens with the dusk outside as animated conversations slide into silence. This is the inaugural presentation of my trip; my first attempt to capture the best and worst moments in the eight countries traversed by *The World*'s globe-girdler and me. Like Nellie when she forced her belongings into the small gripsack, I am obliged to squeeze her seventy-two day journey into a forty-five minute sprint. 'Fasten your seatbelts, hold onto your chairs,' I warn my audience. 'We are about to race round the world with globetrotting Nellie Bly.' From her anxious departure in Hoboken, New Jersey to her triumphant arrival back in Jersey City, I gallop my audience through a slideshow of our global travels. The next time I deliver this presentation it will be in London, and it will definitely feature Nellie's hometowns of Apollo and Cochran's Mills.

Someone in the corner flicks a switch and the hall is illuminated. Appreciative applause, more than just polite, sounds for this Londoner who has come to tell them about their own hometown girl. I am accepted into their fold. As we file out, society member Donna Darlene Dunmore greets me with an A3, hand-stencilled, flamingo pink poster celebrating her hometown's bicentennial in two years' time. It is for Her Majesty the Queen, she explains. There is nothing to do but say I will try. I roll it up and carry it back to England, but the poster never gets past the gates of Buckingham Palace.

Downstairs in the AAHS museum, the coffee we could smell earlier is being poured, along with jugs of iced tea. Home-baked goodies piled neatly on platters vie for space on a conference table – chocolate chip, oatmeal and peanut butter cookies, lemon meringue pie, cupcakes heaped with butter cream frosting, rice crispy bars, fudge brownies. It is a cornucopia of American delights with no regard for calories or sugar content. There are no healthy options. We indulge.

For me the best treat of all is a floor-to-ceiling glass display case entirely devoted to Nellie Bly memorabilia – a shrine. Inside an autographed copy of Nellie's book *Ten Days in a Mad-House* is held together by string, its dog-eared pages bronzed with time. Nearby, on a 3 x 5 inch trading card, Nellie is balancing on a tightrope between stars and planets, dressed in her plaid ulster coat and clutching her gripsack to advertise Schnenk's Mandrake pills which cure 'all bilious and liver complaints'. Another shelf holds a hand-painted porcelain Nellie Bly doll standing six inches high, next to a framed certificate of its authenticity from the United States Historical Society. Nellie is eleventh in a series of American Women of Arts and Letters. 'This Nellie Bly collector doll, created by the artists of the United States Historical Society Doll Shop,' the certificate reads, 'is a tribute to the trailblazing journalist/adventurer and all the women who have followed her footsteps in American journalism.' The historical society has bagged number 841 from an issue of 9,500 Nellies.

It feels good to be in the company of true, blue Nellie Bly fans. They know as much, and often more than I do, about her. Nothing needs to be explained here as it did in Sri Lanka, Singapore, Hong Kong, China, Japan, the UK, even elsewhere in the USA. Except for Dolly, I have never met anyone here tonight before, yet it feels as if they have always been a part of my life. They can practically end my sentences for me. The folks of Apollo have been keeping Nellie Bly's memory alive for decades … and they seem pleased as punch that I have honoured their favourite daughter.

Across thirty-three days, 22,500 miles, six international flights, many trains, buses and now a rental car to travel here, I have finished up where Nellie Bly – Elizabeth Jane Cochran – started out. Nellie and I have shared a challenging, yet exhilarating, voyage around the world and across 125 years. Here in Apollo, I get the sense that my journey is now complete. What started out as Nellie Bly's story is now a part of my own.

Appendix 1

Nellie Bly's Life

1864 – Born Elizabeth Jane Cochran on 5 May in Cochran's Mills, Pennsylvania.

1869 – Moves to Apollo, Pennsylvania.

1871 – Death of father Michael Cochran on 19 July.

1879 – Enrols at Indiana State Normal School to become a teacher in September; unable to pay fees and forced to drop out in December.

1880 – Moves to Allegheny City, Pennsylvania.

1885 – First article 'The Girl Puzzle' published in *The Pittsburg Dispatch*; hired as staff reporter.

1886 – Travels to Mexico as a foreign correspondent in February; returns to Pittsburg in June.

1887 – Moves to New York City in May. Receives first assignment from *The New York World* on 22 September: feigns madness to enter New York City Lunatic Asylum on 24 September; released on 4 October. *Behind Asylum Bars* published on 9 October; *Inside the Madhouse* published on 16 October. *Ten Days in a Mad-House* published in December.

1888 – *Six Months in Mexico* published in January. Proposes assignment to circle the world in less than eighty days to *The New York World* in November.

1889 – Given three days' notice to begin race around the world on 11 November: departs from Hoboken, New Jersey on SS *Augusta Victoria* at 9.40 am on 14 November.

1890 – Breaks world record for circling the world on 25 January at 3.51 pm, arriving back in Jersey City seventy-two days after setting off from Hoboken. Leaves *The New York World* and begins a forty-city lecture tour in February which is abandoned in May. *Around the World in Seventy-Two Days* published in July.

1893 – Returns to journalism at *The New York World*.

1895 – Marries millionaire industrialist Robert Livingstone Seaman, aged 70, on 5 April.

1896–1899 – Spends three years in Europe with her husband.

1899 – Named president of her husband's firm Iron Clad Manufacturing Company in Brooklyn and assumes control of business. Becomes one of America's leading women industrialists. Institutes fair pay and well-being benefits for workers, streamlines production methods and patents several inventions for the prospering company.

1904 – Husband Robert Seaman dies on 11 March.

1911 – Iron Clad Manufacturing Company forced to file for bankruptcy after company funds embezzled by employees.

1912 – Returns to journalism at *The New York Evening Journal*.

1913 – Rides horseback as herald for the first Woman Suffrage Procession on 3 March.

1914 – Leaves America for Austria. Becomes first woman reporting from the Eastern Front – Serbia and Austria – in the First World War.

1919 – Returns to America. Writes for *The New York Evening Journal*; runs newspaper campaigns for poor people and abandoned children.

1922 – Dies 27 January. Buried in Woodlawn Cemetery, New York City.

1978 – Headstone at Woodlawn Cemetery dedicated by New York Press Club on 22 June.

1998 – Introduced into National Women's Hall of Fame, Washington DC.

2002 – Commemorated on US postage stamp.

2015 – 151st birthday celebrated with musical Google Doodle on 5 May.

2019 – Announcement of memorial installation, *The Girl Puzzle*, to be created in her honour on Roosevelt Island, New York City by artist Amanda Matthews of Prometheus Art.

2020 – Memorial statue erected at Pittsburgh International Airport.

Appendix 2

Nellie's Round the World Itinerary

14 November 1889–25 January 1890

14–22 November: SS *Augusta Victoria* from Hoboken, New Jersey to Southampton, England.

22 November: Mail train from Southampton to London, then to Folkestone for ferry to Boulogne, France. Train to Amiens to visit Jules and Honorine Verne, then to Calais.

23–25 November: Mail train from Calais to Brindisi, Italy.

25 November–8 December: SS *Victoria* from Brindisi to Colombo, Ceylon via Port Said, Egypt, Suez Canal, and Aden (Yemen).

8–14 December: Colombo, Ceylon. Day trip to Kandy by train.

14–17 December: SS *Oriental* from Colombo to Singapore via Penang, Malaya.

18 December: Singapore.

18–23 December: SS *Oriental* from Singapore to Hong Kong.

23–24 December: Hong Kong.

24–25 December: SS *Powan* round trip to Canton (Guangzhou), China for Christmas Day.

26–28 December: Hong Kong.

28 December–2 January: SS *Oceanic* from Hong Kong to Yokohama, Japan.

2–7 January: Yokohama. Visits to Tokyo and Kamakura.

7–21 January: SS *Oceanic* from Yokohama to San Francisco.

21–25 January: Trains from San Francisco to Jersey City via Chicago, Pittsburgh and Philadelphia.

25 January: Arrived Jersey City after travelling 21,740 miles in 72 days, 6 hours and 11 minutes. Celebratory parade by ferry across the Hudson River into Manhattan, New York City.

Rosemary's Round the World Itinerary

6 September–8 October 2014

Pre-trip:
London – To follow Nellie Bly's footsteps across my own city.
Amiens, France – To visit Jules Verne's house.
Calais and Boulogne, France.

6–7 September: Sri Lankan Airlines flight from London to Colombo, Sri Lanka.

7–9 September: Colombo.

9–10 September: Round trip by train to Kandy.

11 September: Sri Lankan Airlines flight from Colombo to Singapore.

11–14 September: Singapore.

14 September: Cathay Pacific flight from Singapore to Hong Kong.

14–16 September: Hong Kong.

16–18 September: Round trip by train to Canton (Guangzhou), China.

18–19 September: Hong Kong.

19 September: Cathay Pacific flight from Hong Kong to Tokyo, Japan.

19–21 September: Tokyo. Day trip by train to Yokohama.

21–23 September: Hakone by train and car. Visit to Kamakura.

23–24 September: Tokyo.

24 September: Japan Airlines flight from Tokyo to New York City.

24–29 September: Manhattan. Visits to Hoboken and Jersey City, Roosevelt Island and Woodlawn Cemetery.

29 September: American Airlines flight from New York City to Washington, DC.

29 September–5 October: Washington, DC.

5 October: By car from Washington, DC to Apollo, Pennsylvania.

5–7 October: Apollo. Visit to Cochran's Mills.

7 October: By car from Apollo to Washington, DC.

7–8 October: British Airways flight from Washington, DC to London.

8 October: Arrived in London after travelling 22,500 miles in 33 days.

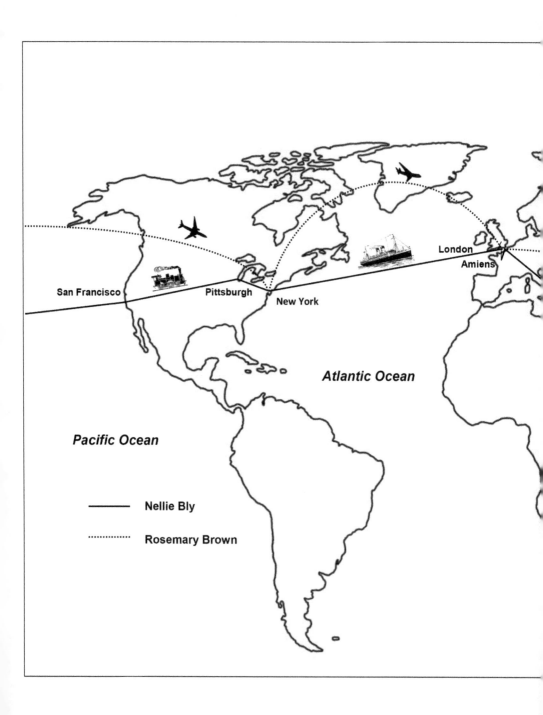

San Francisco

Pittsburgh

New York

London

Amiens

Atlantic Ocean

Pacific Ocean

—————— Nellie Bly

················ Rosemary Brown

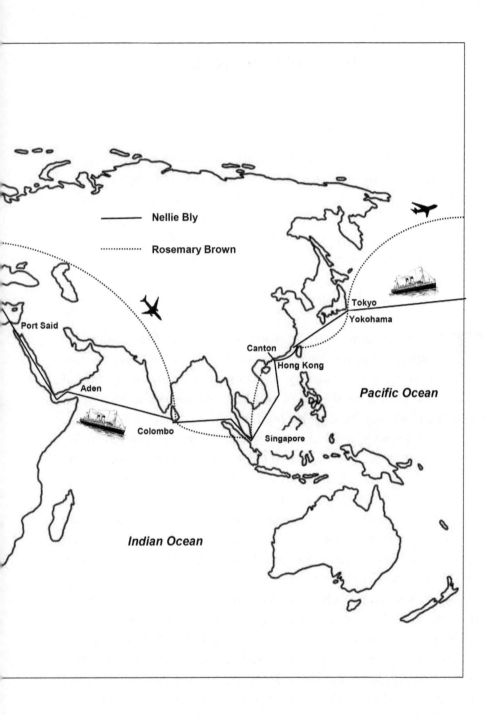

Nellie Bly

Rosemary Brown

Port Said

Aden

Colombo

Singapore

Canton

Hong Kong

Tokyo

Yokohama

Pacific Ocean

Indian Ocean

Acknowledgements

'To so many people this wide world over am I indebted for kindnesses that I cannot, in a little book like this, thank them all individually. They form a chain around the earth. To each and all of you, men, women and children, in my land and in the lands I visited, I am most truly grateful. Every kind act and thought, if but an unuttered wish, a cheer, a tiny flower, is imbedded in my memory as one of the pleasant things of my novel tour.'

Nellie Bly, *Around the World in Seventy-Two Days*, 1890

Thank you to my own 'chain around the earth'. The 'kindnesses' that enriched my travels did not stop there. All along the way I have been inspired by the enthusiasm, encouragement and guidance of friends, family and those I have encountered in my travels, research and writing. To all of my loyal friends in the UK, USA, France and Ireland, I send heartfelt thanks for believing in me ... and in Nellie. I am grateful to the librarians at the Royal Geographical Society, the London Library, and the British Library where happily *The New York World* is on microfilm. I am indebted to the Royal Literary Fellowship Scheme which provided me with hours of priceless one-on-one advice from leading authors: Jennifer Potter, Lynn Knight, Max Eilenberg and especially Wendy Moore. Writer and teacher Wes Brown at London's City Lit was also a guiding light.

I owe a considerable debt to Brooke Kroeger's *Nellie Bly: Daredevil, Reporter, Feminist* and Matthew Goodman's *Eighty Days*. These remarkable books enhanced my travels, my research and my writing.

Thank you to those who joined the 'Nellie Bly trail team' en route. In Sri Lanka: Lakmini, Jévon and Devin Raymond, Jagdesh Mirchandani, and Stephanie and Moahan Balendra. In Japan: Yoshihisa and Yoshie Togo. In New York City: Alice Robbins-Fox and Sally Emery. In Pennsylvania:

Dolly Lackey McCoy and Arnold Blystone. In Washington, DC: Louisa Peat O'Neil.

What more could a writer wish for than a quiet space to get down to work, especially if it is in Paris and Stratford-upon-Avon. Thank you, Patricia Streifel and Margaret Cund, for the luxury of distraction-free writing time in your lovely flats.

To the two who stood by me daily from the dream of the journey to the reality of the book, I express my love and gratitude. Acadia, my daughter, thank you for your Nellie Bly-style fortitude and insistence on the occasional happy hour. It is indeed ironic that after all of this writing, I cannot find the words to thank the book's greatest ally and artist of the illustrations that adorn it – my husband David Stanton. With me step by step from finalising the travel itinerary to proof-reading the last page, suffice it to say that there would be no book without him.

Rosemary J. Brown
London

Author's Note

In writing this book I have drawn primarily on Nellie Bly's book *Around the World in Seventy-Two Days* (1890) and newspaper accounts from *The New York World* and *The Evening World* in 1889 and 1890. As this book is a travelogue, not an academic work, I have chosen to limit footnotes. Works consulted are listed in the Select Bibliography.

In the text, I use the names of locations based on Nellie Bly's time, when Pittsburgh was Pittsburg, Sri Lanka was Ceylon and Guangzhou was Canton. My chapter titles begin with 'In Which Nellie…' as a tribute to the form Jules Verne used in *Around the World in Eighty Days*.

This account was compiled during and after a world journey to celebrate the 125th anniversary of Nellie Bly's epic expedition. I hope readers will be inspired by her remarkable ability to achieve the impossible, her sense of justice and her humanitarianism. In reading this book, perhaps your own sense of adventure will be ignited along with an enhanced belief that 'nothing is impossible if one applies a certain amount of energy in the right direction'.

To offset carbon emissions from my Nellie Bly expedition, I made a contribution to the Rainforest Foundation UK. As a writer there, I learned first-hand the vital role that rainforests play in countering global warming. Thanks to the generosity of friends, I was able to raise money during my travels for the work of UNICEF-UK to commemorate Nellie Bly and her tireless campaigns on behalf of disadvantaged mothers and children.

Chapter Notes

Chapter 1

1. *The New York Times*, 28 January, 1893.

Chapter 2

1. Brooke Kroeger. *Nellie Bly: Daredevil, Reporter, Feminist*. New York: Times Books, 1994, p. 145.

Chapter 3

1. Thomas Whiteside, *The Tunnel Under the Channel*, London: Simon & Schuster, 1962, pp. 17, 25.
2. Robert Harborough Sherard. *Twenty Years in Paris: Being Some Recollections of a Literary Life*. London: Hutchinson & Co, 1905, p. 314.

Chapter 6

1. www.submerged.co.uk, www.shipsproject.org/Wrecks/Wk_Nepaul.html

Chapter 8

1. Noël Coward. *The Complete Illustrated Lyrics*. New York: Harry N Abrams, 1998.

Chapter 11

1. John Russell Young. *Around the World with General Grant, Vol. 2*. New York: The American News Company, 1879, p. 482.
2. John Y. Simon. *The Papers of Ulysses S. Grant, Volume 29: October 1, 1878– September 30, 1880*. Carbondale, IL: Southern Illinois University Press, 2008, p. 228.
3. Ibid. p. 193.

Chapter 12

1. Charles Dickens. *American Notes for General Circulation*. London: Chapman and Hall, 1842.
2. Nellie Bly, *Ten Days in a Mad-House*, New York: Munro, 1887, p. 29.

Chapter 14

1. Until 1911, Pittsburgh was sometimes spelled Pittsburg, as in *The Pittsburg Dispatch*.
2. Nellie Bly's *nom de plume* was a misspelling of Stephen Foster's minstrel song *Nelly Bly*, published in 1850.
3. Brooke Kroeger. *Nellie Bly: Daredevil, Reporter, Feminist*. New York: Times Books, 1994, p. 13.

Select Bibliography

Bly, Nellie. *Ten Days in a Mad-House*. New York: Munro, 1887.

Bly, Nellie. *Among the Mad*. Godey's Lady's Book. Vol 118. January 1, 1889.

Bly, Nellie. *Nellie Bly's Book: Around the World in Seventy-Two Days*. New York: Pictorial Weeklies, 1890.

Davidson, Lillias Campbell. *Hints to Lady Travellers: At Home and Abroad*, reprint edition for Royal Geographical Society. London: Elliott and Davidson Ltd, 2011.

De Blaquiére, Dora. *The Purchase of Outfits for India and the Colonies*. Girl's Own Paper, 1890 Annual.

Goodman, Matthew. *Eighty Days: Nellie Bly and Elizabeth Bisland's History-Making Race Around the World*. New York: Ballentine Books, 2013.

Gray, John Henry. *Walks in the City of Canton*. Hong Kong: De Souza, 1875.

Hahn, Emily. *Around the World with Nellie Bly*. Boston: Houghton Mifflin, 1959.

Henry, T. J. *1816–1916; History of Apollo, Pennsylvania: The Year of a Hundred Years*. Apollo, Pennsylvania: News-Record Publishing Company, 1916.

Kroeger, Brooke. *Nellie Bly: Daredevil, Reporter, Feminist*. New York: Times Books, 1994.

Matthews, Mimi. *A Victorian Lady's Guide to Fashion and Beauty*. Barnsley: Pen & Sword Books, 2018.

McDougall, Walt. *This is the Life!* New York: Alfred A. Knopf, 1926.

Palin, Michael. *Around the World in 80 Days*. London: BBC Books, 1989.

Rittenhouse, Mignon. *The Amazing Nellie Bly*. New York: E P Dutton, 1956.

Sherard, Robert Harborough. *Twenty Years in Paris: Being Some Recollections of a Literary Life*. London: Hutchinson & Co, 1905.

Stoddard, John L. *Glimpses of the World; a Portfolio of Photographs of the Marvelous Works of God and Man*. Chicago: R.S. Peale Company, 1892.

Verne, Jules. *Around the World in Eighty Days* (1873). Translated by William Butcher. London: Oxford University Press, 2008.

Whiteside, Thomas. *The Tunnel Under the Channel*. London: Simon & Schuster, 1962.

Index

Plates are indicated in **bold**; page numbers of illustrations are indicated in *italic*.